NAVIGATING THE BUSINESS SIDE OF BIOMEDICINE AND BIOTECHNOLOGY

RASIT DINC

Navigating the Business Side of Biomedicine and Biotechnology
Rasit Dinc (Author)

ISBN: 978-83-67405-76-8
e-ISBN: 978-83-67405-77-5

sciendo

Published by Sciendo
Part of Walter de Gruyter GmbH, Berlin
Published with paid access at www.sciendo.com

NAVIGATING THE BUSINESS SIDE OF BIOMEDICINE AND BIOTECHNOLOGY

CONTENTS

TABLE OF FIGURES

PREFACE

The pages that follow embarks on an explorative journey into the intricate nexus of biotechnology and finance by delving into the synergistic interplay between groundbreaking scientific advancements and the complex dynamics of economic frameworks. It is an odyssey that charts how scientific discoveries are transformed into economic assets and how financial models become the catalysts for biomedical innovation.

As we traverse the landscape where science and finance converge, we navigate a path lined with both risks and opportunities. Herein, we unravel the process by which scientific breakthroughs are translated into financial resources, and how financial strategies are ingeniously woven into the fabric of scientific progress.

Throughout this book, we will span a diverse array of topics, from the intricacies of Monte Carlo simulations to the nuances of option pricing models, and from drug development for rare diseases to sophisticated investment strategies.

The purpose of this work is twofold: to elucidate the financial dimensions of scientific research for the scholarly reader, and to provide financial professionals with insights into the valuation of such endeavors. In this discourse, we will discuss how the economic value of scientific innovation can be maximized and how financial instruments can be utilized more effectively.

Each chapter is an invitation to explore the challenges and opportunities resonating within this unique intersection of science and finance. By the conclusion of this book, readers are expected to gain a deeper understanding of how these two realms intertwine and influence each other. This work aims to bridge the worlds of science and finance, opening new horizons in both fields.

EXPLORING THE NATURE OF HEALTHCARE FINANCE

Healthcare finance encompasses a broad spectrum of topics, each intertwined with the intricate fabric of healthcare delivery and policy. The forthcoming section aims to elucidate these topics in a manner that is both accessible and academically rigorous.

We will delve into the fundamental principles of health economics, the various models of financing healthcare, and the pivotal role of financial management within health institutions. Moreover, the intersection of healthcare financing with policy-making will be scrutinized, highlighting how financial considerations shape healthcare outcomes and accessibility.

The purpose of this introductory segment is twofold. Firstly, it seeks to demystify the complexities of healthcare finance for those new to the subject, laying a clear and comprehensive groundwork. Secondly, for those already familiar with the rudiments of finance or healthcare, it offers a refined lens through which to examine these disciplines in concert.

As we embark on this journey, readers are encouraged to approach the material with an open and inquisitive mind. The nuances of healthcare finance are not only pivotal in understanding the broader healthcare landscape but also essential in forging effective strategies and policies in this sector. Whether you are a healthcare professional, a financial expert, or an interested observer, the insights gained here will be invaluable in navigating the multifaceted world of healthcare finance.

Biotechnology and drug development have revolutionized the healthcare industry, offering cutting-edge solutions to previously untreatable conditions. This dynamic field not only holds the promise of breakthrough medical treatments but also presents lucrative opportunities for investors and financial institutions.

The intersection of healthcare and finance is a multifaceted and ever-evolving landscape. Biotechnology and drug development not only have the potential to transform the treatment of diseases but also offer a glimpse into the future of healthcare investment and financial markets. Understanding the complexities of this sector requires a comprehensive examination of the scientific, regulatory, and economic factors that drive innovation and investment in biotech and drug development.

At the heart of this intersection lies the intricate relationship between scientific discovery and financial feasibility. As new technologies and medical breakthroughs continue to emerge, investors are tasked with evaluating the risks and potential returns associated with funding biotech startups and drug development ventures. This delicate balance between scientific innovation and financial prudence is what shapes the investment landscape in healthcare and biotechnology.

In order to navigate this terrain, it is essential to explore the regulatory frameworks that govern drug development, the pivotal role of clinical trials in determining a drug's efficacy and safety, and the market dynamics that influence the commercialization and pricing of innovative treatments. Furthermore, understanding the impact of healthcare policies, reimbursement models, and ethical considerations on the financial viability of biotech companies is crucial for investors and stakeholders in the healthcare and finance sectors.

While it is undeniable that biotechnology and drug development have the potential to revolutionize the healthcare industry and offer lucrative opportunities for investors, it is important to acknowledge the significant risks and uncertainties associated with investing in this sector. The complexities of drug development, including high failure rates in clinical trials and lengthy regulatory approval processes, present substantial challenges for investors seeking predictable returns on their investment.

The biotechnology and drug development landscape are characterized by intense competition, rapidly changing scientific advancements, and the constant need for substantial financial resources. The high costs of research and development, coupled with the uncertainty of success, make investments in this sector inherently risky. Moreover, the biotech industry is vulnerable to shifts in regulatory policies, market dynamics, and public perception, which can significantly impact the financial viability of companies operating in this space.

It is crucial for investors to carefully consider the inherent risks and uncertainties associated with biotech and drug development. While the potential for groundbreaking medical advancements and significant financial returns exists, the volatile nature of this industry demands a cautious and discerning approach to investment. Understanding the intricacies of scientific, regulatory, and economic factors is essential for making well-informed investment decisions and mitigating the inherent risks posed by the convergence of healthcare and finance.

Regulatory Oversight and Clinical Trials

Navigating the terrain of biotechnology and drug development involves an in-depth understanding of the regulatory frameworks that govern the industry. Regulatory approval processes, particularly the pivotal role of clinical trials in determining a drug's efficacy and safety, significantly influence the feasibility of investments in this sector. Investors must carefully evaluate the regulatory landscape to gauge the impact of compliance and approval timelines on investment outcomes.

Financial Strategy and Market Dynamics

The intricate relationship between scientific innovation and financial prudence permeates the landscape of healthcare and biotechnology. Market dynamics, reimbursement models, and ethical considerations intersect with financial strategy, shaping the commercialization and pricing of innovative treatments. Understanding these economic factors is paramount for investors and stakeholders seeking to navigate the complexities of this dynamic landscape.

Risks and Opportunities

While the landscape of biotechnology and drug development presents compelling opportunities for scientific and financial exploration, it is crucial to recognize the inherent risks and uncertainties embedded within this sector. High failure rates in clinical trials, the volatility of regulatory policies, and the substantial financial resources required for research and development underscore the unpredictable nature of investments in biotech and drug development.

Strategic Decision-Making

Given the intricate relationship between healthcare and finance, investors and stakeholders must approach this convergence with meticulous consideration. Strategic decision-making demands a comprehensive understanding of the interplay between scientific progress, regulatory oversight, and financial viability. By critically evaluating the landscape of biotechnology and drug development, informed decisions can be made to mitigate risks and capitalize on opportunities in this ever-evolving industry.

The exploration of this dynamic intersection offers a profound insight into the captivating world of biotechnology and drug development, shedding light on the opportunities and challenges that define this complex landscape. As we continue to navigate the complexities of this convergence, it becomes increasingly evident that the future of healthcare and finance is intricately intertwined, presenting a compelling yet challenging terrain for those involved.

Bridging the Knowledge Gap between Science and Commerce

In today's rapidly evolving healthcare landscape, bridging the knowledge gap between science and commerce is more important than ever. As biotechnology and drug development continue to advance at a rapid pace, there is a growing need for individuals who can effectively navigate both the scientific and financial aspects of these fields.

While scientists and researchers play a crucial role in developing innovative therapies and treatments, financial decision-makers are responsible for ensuring the sustainability and success of these endeavors.

This bridge between science and commerce is critical for ensuring that groundbreaking research and innovative treatments can successfully transition from the laboratory to the market, where they can ultimately benefit patients. It also plays a pivotal role in attracting and retaining investment in biotech and drug development, as investors seek assurance that their financial resources will be effectively utilized in driving scientific progress and achieving tangible healthcare outcomes.

Understanding the complexities involved in healthcare finance and the unique challenges posed by biotechnology and drug development is essential for all stakeholders. As such, this document aims to provide a comprehensive overview of the key considerations, trends, and strategies at the intersection of science and finance. By shedding light on this intricate relationship, stakeholders can make informed decisions and contribute to the advancement of healthcare through sound financial management and investment practices.

The Impact of Healthcare Financing on Biotechnology and Drug Development

As the impact of healthcare financing on biotechnology and drug development continues to gain prominence, it is crucial to understand the intricate dynamics at play. Healthcare financing not only influences the pace of innovation and research in the biotech and pharmaceutical sectors but also shapes the accessibility of groundbreaking treatments to patients.

The availability of funding directly impacts the research and development efforts of biotech and pharmaceutical companies. It determines the resources allocated to the discovery and testing of new drugs, as well as the pursuit of novel therapeutic approaches. In turn, this influences the pipeline of potential treatments that could address unmet medical needs and improve patient outcomes.

Moreover, the relationship between healthcare financing and biotechnology extends to the regulatory landscape. Financial considerations often intersect with regulatory requirements, shaping the pathways for drug approval, patient access, and reimbursement. Understanding these interactions is essential for both healthcare and finance professionals, as it enables them to anticipate and navigate the complexities of bringing innovative therapies to market.

In the realm of investment, the impact of healthcare financing on biotechnology and drug development is multifaceted. Investors play a pivotal role in allocating capital to companies engaged in biotech and drug development. The flow of investments not only fuels research initiatives but also influences the valuation and commercialization prospects of these companies. In this context, evaluating the financial viability and potential return on investment in the biotech and pharmaceutical sectors becomes paramount.

Continued awareness of the intricate interplay between healthcare financing and biotechnology is essential for stakeholders across the healthcare and finance domains. By seamlessly integrating the knowledge of healthcare finance with the dynamic landscape of biotech and drug development, stakeholders can effectively shape the future of healthcare innovation and investment.

Strategies for Successful Healthcare Financing in Biotech and Drug Development

In order to navigate the complex landscape of healthcare financing in biotechnology and drug development, stakeholders must employ strategic approaches that align scientific innovation with financial sustainability. One such strategy involves fostering collaborations between scientific and financial experts to ensure that research and development efforts are supported by adequate funding and strategic investment.

Furthermore, stakeholders should proactively stay abreast of evolving regulatory frameworks and reimbursement dynamics to anticipate the impact of healthcare financing on the commercialization of innovative treatments. By understanding the intersection of financial considerations and regulatory requirements, organizations can position themselves for success in bringing novel therapies to market.

Another critical aspect of successful healthcare financing in biotech and drug development is the engagement of informed investors who recognize the value of long-term investment in pioneering medical advancements. Educating investors about the complexities and potential of biotechnology and drug development can foster a supportive investment environment that facilitates sustainable growth and innovation in the healthcare sector.

Moreover, the integration of healthcare finance expertise into the decision-making processes of biotech and pharmaceutical companies is essential for optimizing resource allocation and ensuring financial prudence throughout the drug development lifecycle. By implementing robust financial management practices, organizations can mitigate risks and enhance the efficiency of research and development initiatives.

By adopting these strategic approaches, stakeholders can effectively bridge the gap between healthcare and finance, fostering a synergistic relationship that drives groundbreaking scientific progress while securing sustainable financial backing. This holistic approach is integral to accelerating the translation of innovative medical discoveries into tangible benefits for patients and maximizing the potential of biotechnology and drug development.

The Role of Collaboration in Bridging the Knowledge Gap

Collaboration between scientific and financial experts is paramount in bridging the knowledge gap between science and commerce in the healthcare industry. By fostering a collaborative environment, organizations can leverage the expertise of both domains to make informed decisions that drive innovation while ensuring financial sustainability.

In this collaborative model, scientific experts provide insights into the latest advancements in biotechnology and drug development, allowing financial professionals to assess the potential impact on healthcare outcomes and market viability. Conversely, financial experts offer valuable perspectives on investment strategies, risk assessment, and financial forecasting, enabling scientific teams to make strategic decisions that align with market and industry dynamics.

By breaking down silos and promoting open communication, organizations can maximize the synergies between scientific innovation and financial prudence. This collaborative approach not only optimizes resource allocation but also ensures that research and development efforts are aligned with the broader financial goals of the organization.

Leveraging Data Analytics for Informed Decision-Making

Data analytics plays a pivotal role in bridging the knowledge gap between science and finance in the healthcare sector. By harnessing the power of data, organizations can gain actionable insights into market trends, patient demographics, resource allocation patterns, and financial performance metrics.

Utilizing advanced analytics tools, stakeholders can assess the financial implications of different research and development pathways, identify opportunities for cost optimization, and forecast the potential return on investment for innovative therapies. At the intersection of science and finance, data analytics enables evidence-based decision-making that mitigates risks and enhances the efficiency of healthcare financing in biotech and drug development.

Moreover, leveraging data analytics allows organizations to track the impact of healthcare financing on research outcomes, patient access, and market penetration of new treatments. This insight is invaluable for shaping investment strategies, refining commercialization plans, and ensuring that financial resources are allocated to initiatives with the greatest potential for patient benefit.

Embracing Technological Innovation for Financial Management

The integration of technological innovation in financial management is pivotal for bridging the knowledge gap between science and finance in healthcare. From advanced financial modeling tools to blockchain-based transaction systems, technology offers opportunities to streamline financial processes, enhance transparency, and mitigate operational risks in biotechnology and drug development.

By embracing technological innovation, organizations can establish robust financial management frameworks that align with the dynamic nature of scientific research and development. Automated reporting systems, real-time financial data analytics, and secure transaction platforms empower stakeholders to make agile financial decisions that support the ever-evolving landscape of healthcare innovation.

In conclusion, the continued collaboration between scientific and financial experts, the strategic utilization of data analytics, and the integration of technological innovation are imperative for bridging the knowledge gap between science and commerce in the context of healthcare financing. This multifaceted approach empowers stakeholders to navigate the complexities of healthcare financing while driving innovation and ensuring the sustainable advancement of biotechnology and drug development.

Implementing Ethical and Sustainable Financial Practices

In addition to the strategic approaches outlined, it is crucial for stakeholders in biotechnology and drug development to prioritize ethical and sustainable financial practices. This involves integrating principles of corporate social responsibility and ethical investment into the fabric of financial decision-making processes.

Ethical considerations should extend to the allocation of financial resources, ensuring that investments are channeled towards research and development endeavors that align with ethical standards and hold the potential to address unmet medical needs. Furthermore, sustainable financial practices should encompass long-term planning that accounts for the environmental and societal impact of healthcare innovations, thereby promoting responsible innovation within the industry.

Adopting ethical and sustainable financial practices not only fosters trust and credibility within the healthcare sector but also positions organizations as responsible stewards of financial resources, thereby attracting investors who prioritize ethical investment and sustainability.

By balancing financial sustainability with ethical considerations, stakeholders can cultivate a culture of responsible financial stewardship, ultimately contributing to the long-term viability and positive societal impact of biotechnology and drug development.

Importance of Economic Understanding in Healthcare

Understanding the fundamentals of healthcare economics is paramount for stakeholders in the biotechnology and drug development industry. It provides a crucial framework for evaluating the economic impact of healthcare innovations, making informed investment decisions, and ensuring the efficient allocation of financial resources.

At its core, economic understanding in healthcare enables stakeholders to assess the cost-effectiveness of new therapies, analyze the economic burden of diseases, and evaluate the return on investment for research and development initiatives. By integrating economic principles into decision-making processes, organizations can optimize resource allocation, maximize the value of healthcare interventions, and drive sustainable financial outcomes.

Leveraging Economic Principles for Decision-Making

The application of economic principles in healthcare drives evidence-based decision-making that aligns with both financial sustainability and patient-centric outcomes. Stakeholders can utilize economic analyses to compare the cost and benefits of different treatment modalities, assess the long-term economic implications of healthcare innovations, and navigate pricing strategies in a competitive market landscape.

Moreover, understanding healthcare economics empowers stakeholders to address healthcare disparities, optimize access to innovative therapies, and contribute to the broader goal of improving public health outcomes while maintaining financial prudence. This holistic approach considers not only the financial implications of healthcare interventions but also their societal and economic impact, fostering a balanced approach to decision-making in biotechnology and drug development.

In conclusion, a comprehensive understanding of healthcare economics is essential for stakeholders in the biotechnology and drug development sector. By leveraging economic principles, organizations can navigate complex market dynamics, optimize resource allocation, and drive sustainable financial practices while advancing the frontiers of healthcare innovation.

Understanding the fundamentals of healthcare economics is crucial for stakeholders in biotechnology and drug development. A solid grasp of economic principles in the context of healthcare enables organizations to make informed decisions that drive innovation, ensure patient access to groundbreaking therapies, and optimize the allocation of financial resources.

Cost-Benefit Analysis and Resource Allocation

Healthcare economics involves conducting cost-benefit analyses to evaluate the potential impact of new therapies on patient outcomes and healthcare expenditure. By assessing the cost-effectiveness of different interventions, organizations can prioritize resource allocation to maximize patient benefit while maintaining financial sustainability.

Market Dynamics and Access to Innovation

An understanding of healthcare economics also equips stakeholders to navigate the complex market dynamics that influence the adoption and access of innovative therapies. By analyzing factors such as reimbursement policies, pricing strategies, and market competition, organizations can tailor their commercialization efforts to ensure the widespread availability of transformative treatments.

Policy and Regulatory Considerations

Additionally, healthcare economics encompasses an understanding of healthcare policy and regulations, which play a pivotal role in shaping the financial landscape of biotechnology and drug development. By staying abreast of evolving regulatory frameworks and policy changes, organizations can adapt their financial strategies to align with the changing healthcare environment.

In essence, a strong foundation in healthcare economics empowers stakeholders to make strategic financial decisions that balance patient impact with economic viability, ultimately driving sustainable innovation in biotechnology and drug development.

The Impact of Healthcare Economics on Biotechnology and Drug Development

Understanding the fundamentals of healthcare economics is essential for stakeholders in biotechnology and drug development to navigate the complex landscape of healthcare financing. Healthcare economics encompasses the study of how healthcare resources are allocated and the impact of these decisions on patient outcomes, cost-effectiveness, and overall healthcare system performance.

In the context of biotechnology and drug development, a thorough understanding of healthcare economics enables stakeholders to evaluate the cost-effectiveness of new therapies, assess the potential market demand, and make informed decisions about resource allocation. By analyzing healthcare economics, organizations can identify opportunities to optimize research and development investments, maximize patient access to innovative treatments, and contribute to the long-term sustainability of healthcare systems.

Furthermore, an understanding of healthcare economics empowers stakeholders to engage in evidence-based discussions with payers, policymakers, and regulatory authorities. By articulating the economic value of innovative therapies and the potential impact on healthcare expenditure, organizations can navigate market access challenges and advocate for the adoption of new technologies within healthcare systems.

In conclusion, an in-depth grasp of healthcare economics is integral to the success of biotechnology and drug development endeavors. By leveraging this knowledge, stakeholders can make data-driven decisions, demonstrate the value of their innovations, and contribute to the advancement of healthcare systems that prioritize both patient well-being and financial sustainability.

Implementing Value-Based Pricing Strategies

One of the key aspects of understanding healthcare economics in the context of biotechnology and drug development is the implementation of value-based pricing strategies. This approach involves aligning the price of a treatment with the value it provides to patients and the healthcare system.

By incorporating value-based pricing, stakeholders can ensure that the cost of new therapies reflects their clinical benefits, potential to improve patient outcomes, and overall impact on the healthcare system. This not only promotes the responsible use of financial resources but also incentivizes the development of innovative treatments that address unmet medical needs.

Furthermore, understanding healthcare economics and embracing value-based pricing enables organizations to engage in constructive dialogues with payers and reimbursement authorities. By demonstrating the value proposition of their therapies, stakeholders can negotiate pricing that reflects the real-world benefits of their innovations, ultimately enhancing patient access and market penetration.

In summary, integrating value-based pricing strategies based on a solid understanding of healthcare economics is crucial for ensuring the sustainability and affordability of innovative treatments in biotechnology and drug development. This approach facilitates a balance between the financial considerations of stakeholders and the delivery of high-value healthcare solutions to patients.

Embracing Technological Advancements for Financial Management in Healthcare

In today's rapidly evolving healthcare landscape, the integration of technological advancements is key to enhancing financial management practices. Leveraging advanced financial modeling tools and blockchain-based transaction systems enables organizations to streamline processes, improve transparency, and reduce operational risks in biotechnology and drug development financing.

Moreover, the adoption of automated reporting systems and real-time financial data analytics empowers stakeholders to make agile financial decisions that align with the dynamic nature of scientific research and development. These technological solutions not only offer efficiency but also contribute to ensuring the sustainable advancement of biotechnology and drug development by providing valuable insights for financial planning and decision-making.

By embracing technological innovations, organizations can establish robust financial management frameworks that not only bridge the knowledge gap between science and finance in healthcare but also foster an environment conducive to innovation and sustainable growth.

Leveraging Data Analytics for Informed Financial Decision-Making

In addition to technological advancements, leveraging data analytics serves as a critical component in informed financial decision-making in the context of biotechnology and drug development financing. Data analytics provides stakeholders with the ability to identify trends, forecast financial outcomes, and optimize resource allocation to support innovative research and development initiatives.

Utilizing data from diverse sources such as clinical trials, market trends, and patient demographics enables organizations to assess the financial viability of new therapies, anticipate market demand, and tailor financial strategies to maximize patient access while ensuring long-term sustainability.

Furthermore, the integration of data analytics in financial decision-making promotes evidence-based discussions with stakeholders, including investors, regulatory authorities, and payers. This not only fosters a deeper understanding of the economic value of healthcare innovations but also contributes to the development of robust financial strategies that have a positive impact on patient outcomes and overall healthcare system performance.

In conclusion, the synergistic integration of technological advancements and data analytics empowers stakeholders in biotechnology and drug development to navigate the complexities of healthcare financing with informed and strategic financial decision-making.

Prioritizing Ethical and Sustainable Financial Practices

In the realm of biotechnology and drug development, prioritizing ethical and sustainable financial practices is paramount for maintaining integrity and building trust within the healthcare industry. This involves adhering to transparent financial reporting standards, ethical investment practices, and sustainable resource allocation.

Ethical financial practices ensure that stakeholders uphold the highest standards of integrity, accountability, and compliance in all financial transactions and decision-making processes. By prioritizing ethical considerations, organizations build credibility and trust among investors, regulatory bodies, and the broader healthcare community, thereby fostering long-term partnerships and sustainable growth.

Furthermore, embracing sustainable financial practices involves balancing short-term financial gains with long-term societal and environmental impacts. This encompasses responsible stewardship of financial resources, considering the broader implications of financial decisions on public health, environmental conservation, and societal well-being.

In conclusion, by prioritizing ethical and sustainable financial practices, stakeholders in biotechnology and drug development not only uphold their commitment to societal well-being but also contribute to the overall sustainability and trustworthiness of the healthcare industry as a whole.

Beyond technological and analytical considerations, ethical and sustainable financial practices hold a pivotal role in shaping the financial landscape of biotechnology and drug development. Stakeholders must prioritize the integration of ethical principles

Leveraging Financial Insights in Biomedical Research

The intersection of biomedical research and financial decision-making presents a unique opportunity for stakeholders in the healthcare sector to optimize resource allocation and drive innovation. By leveraging financial insights, organizations can strategically prioritize research endeavors, allocate funding to projects with the highest potential for impact, and ensure efficient utilization of financial resources.

Financial data analytics play a pivotal role in identifying trends, evaluating the return on investment in research initiatives, and making informed decisions about future research directions. Through the integration of financial metrics and performance indicators, stakeholders can gain valuable insights into the cost-effectiveness of research projects and make data-driven decisions that align with strategic objectives.

Moreover, the collaboration between biomedical researchers and financial experts facilitates the evaluation of research outcomes in terms of both scientific advancements and financial sustainability. This multidisciplinary approach allows for the identification of research priorities that align with both scientific opportunities and financial viability, ultimately contributing to the efficient utilization of resources and the sustainable advancement of biomedical research.

In conclusion, the symbiotic relationship between biomedical research and financial decision-making empowers stakeholders to optimize resource allocation, drive innovation, and contribute to the long-term sustainability of healthcare advancements. By harnessing financial insights, organizations can align their research efforts with strategic financial goals, thereby maximizing the impact of biomedical research on patient care and healthcare outcomes.

Navigating Financial Decision Making in Biomedical Research

The intersection of biomedical research and financial decision making presents a unique set of challenges and opportunities for stakeholders in the healthcare industry. As organizations pursue scientific advancements in biotechnology and drug development, it is essential to align financial strategies with the dynamic nature of research, innovation, and regulatory requirements.

Leveraging Risk Management Techniques

Integrating risk management techniques into financial decision making is paramount in the context of biomedical research. Given the inherent uncertainties associated with research and development (R&D) activities, stakeholders must adopt robust risk assessment frameworks to evaluate the financial implications of research outcomes, clinical trial results, and regulatory developments.

By leveraging risk management techniques, organizations can proactively identify and mitigate financial risks, optimize resource allocation, and align financial strategies with the risk tolerance levels of stakeholders and investors. This approach fosters resilience in the face of scientific uncertainties, thereby enhancing the financial sustainability of biomedical research initiatives.

Evaluating Funding Options for Research and Development

In the pursuit of biomedical research endeavors, stakeholders are often faced with the challenge of securing adequate funding to support R&D activities. An essential aspect of financial decision making involves evaluating various funding options, including government grants, private investment, and collaboration with philanthropic organizations.

Understanding the nuances of different funding sources, their associated terms, and the impact on intellectual property rights is crucial for making informed financial decisions in biomedical research. Moreover, effective financial management requires organizations to align funding strategies with the long-term research objectives, ensuring continuity and momentum in scientific endeavors.

Embracing Agile Financial Planning

The dynamic and iterative nature of biomedical research necessitates agile financial planning to accommodate evolving research priorities, unexpected findings, and regulatory changes. Traditional financial planning paradigms may not fully capture the adaptive nature of scientific progress, prompting the need for agile budgeting and financial forecasting approaches.

By embracing agile financial planning, organizations can respond swiftly to emerging research opportunities, allocate resources based on real-time insights, and optimize financial strategies to support groundbreaking discoveries. This iterative approach to financial planning aligns with the iterative nature of scientific exploration, enabling stakeholders to navigate the uncertainties of biomedical research with financial agility.

In conclusion, the integration of risk management techniques, strategic evaluation of funding options, and the adoption of agile financial planning are instrumental in navigating the intersection of biomedical research and financial decision making. These approaches empower stakeholders to align financial strategies with scientific priorities, drive innovation, and contribute to the sustainable advancement of biotechnology and drug development.

Role of Finance in Biotech Health Care Delivery

In the realm of biotech health care delivery, the role of finance goes beyond budgeting and financial planning; it encompasses the strategic allocation of resources to ensure the effective delivery of innovative medical solutions to patients. As biotech companies strive to bring cutting-edge therapies and treatments to market, sound financial management plays a critical role in supporting their mission and driving positive health outcomes.

Investment in Patient-Centric Solutions

Finance in biotech health care delivery revolves around allocating resources to develop patient-centric solutions that address unmet medical needs. This includes funding research and development efforts aimed at creating therapies with the potential to significantly improve patient outcomes and quality of life. By strategically investing in these initiatives, biotech companies can contribute to the advancement of medical science while delivering value to patients and healthcare systems.

Balancing Innovation and Affordability

The financial landscape of biotech health care delivery requires careful consideration of the balance between innovation and affordability. While driving scientific breakthroughs is essential, it is equally important to ensure that innovative treatments remain accessible to patients. Financial decision-makers in biotech companies are tasked with navigating this delicate balance, fostering a sustainable approach that supports ongoing innovation while addressing affordability challenges within the healthcare ecosystem.

Leveraging Data-Driven Insights

Informed financial decision-making in biotech health care delivery relies on leveraging data-driven insights to assess the impact of interventions on patient outcomes and healthcare costs. By analyzing real-world evidence and health economic data, finance professionals can contribute to optimizing the value proposition of biotech products and therapies. This approach not only supports strategic pricing and access decisions but also demonstrates the value of biotech innovations in improving health outcomes.

Collaborating for Market Access

Collaboration between finance and market access teams is pivotal in ensuring the successful commercialization of biotech products. Financial expertise is instrumental in developing pricing and reimbursement strategies that facilitate market access while aligning with the needs of patients, providers, and payers. Through close collaboration, finance professionals can contribute to the development of sustainable business models that support the availability of innovative therapies in the healthcare marketplace.

In conclusion, the role of finance in biotech health care delivery extends to shaping the strategic direction of companies, fostering patient-centric innovation, and addressing the economic considerations of novel therapies. By integrating financial expertise with a commitment to improving patient outcomes, biotech companies can navigate the complex landscape of healthcare delivery while driving positive impact and sustainable growth.

Leveraging Financial Instruments for Growth

In addition to strategic allocation of resources, biotech companies can leverage financial instruments to support their growth initiatives. Accessing capital markets through avenues such as initial public offerings, private placements, or venture capital funding can provide the necessary financial backing for research and development, manufacturing capabilities, and commercialization efforts. By judiciously utilizing financial instruments, biotech companies can fuel their expansion while continuing to drive innovation in the healthcare sector.

Integrating Risk Management in Market Expansion

As biotech companies expand their market reach, the integration of risk management practices becomes essential. Market expansion efforts entail a range of financial risks, including currency fluctuations, regulatory compliance challenges in new geographical regions, and market-specific demand variations. By integrating risk management into their expansion strategies, biotech firms can safeguard their financial stability while pursuing growth opportunities in diverse healthcare markets.

Addressing Ethical and Social Responsibility Considerations

Finance professionals in biotech companies play a pivotal role in addressing ethical considerations and social responsibility within funding and investment decisions. Ensuring that financial activities align with ethical standards, responsible sourcing of funding, and considering the social impact of investment decisions are integral components of financial decision-making in the biotech sector. By incorporating these considerations, companies can uphold their commitment to ethical practices and social responsibility while pursuing their financial objectives.

Embracing Sustainable Financial Practices

The integration of sustainable financial practices further reinforces the role of finance in biotech health care delivery. This encompasses practices such as sustainable investment principles, consideration of environmental and social governance factors in financial decision-making, and transparent reporting on financial and non-financial performance. Embracing sustainable financial practices not only aligns with the values of stakeholders and investors but also contributes to the long-term sustainability and resilience of biotech companies in the rapidly evolving healthcare landscape.

In summation, the role of finance in biotech health care delivery extends from supporting innovative research to shaping market access strategies, from leveraging financial instruments for growth to integrating ethical and sustainable practices. By incorporating these facets into financial decision-making, biotech companies can continue to drive positive impact, foster innovation, and contribute to the advancement of healthcare delivery for the benefit of patients and society as a whole.

Fundamentals of Drug Development Economics

Drug development economics encompasses a complex interplay of financial considerations, regulatory requirements, and market dynamics. As biotech companies embark on the journey of bringing innovative therapies to patients, understanding the fundamentals of drug development economics becomes paramount in making strategic financial decisions.

Cost of Research and Development

The cost of research and development constitutes a significant portion of the expenses incurred in bringing a new drug to market. From early-stage discovery to clinical trials and regulatory approvals, biotech firms must meticulously allocate financial resources to support R&D activities. Understanding the cost implications of each phase of drug development is essential for financial planning and investment decision-making, ensuring that adequate funding is available to progress the pipeline of novel therapies.

Pricing and Market Access Strategies

Strategic pricing and market access strategies are integral to the successful commercialization of biotech products. Financial professionals in the biotech sector are tasked with evaluating pricing models, considering healthcare reimbursement systems, and navigating market-specific dynamics to maximize the accessibility of innovative therapies while ensuring sustainable financial returns. These strategies require a nuanced understanding of the economic value proposition of the therapies and their potential impact on healthcare systems and patient outcomes.

Assessing Economic Value and Affordability

Assessing the economic value and affordability of new therapies is a key consideration in drug development economics. Finance professionals collaborate with commercial and market access teams to conduct health economic analyses and cost-effectiveness evaluations, providing insights into the long-term financial implications of integrating the therapies into healthcare pathways. By quantifying the economic value of improved patient outcomes and considering the affordability for healthcare stakeholders, biotech companies can tailor their pricing and access strategies to align with the broader healthcare landscape.

Evaluating Return on Investment

A critical aspect of drug development economics is the evaluation of the return on investment for R&D expenditure. Finance teams analyze the potential revenue streams, market penetration projections, and competitive landscape to gauge the financial viability of advancing specific drug candidates. Understanding the potential risks and rewards associated with investment in drug development endeavors is essential for optimizing resource allocation and prioritizing projects with the greatest potential for financial and clinical success.

In conclusion, mastering the fundamentals of drug development economics equips finance professionals in the biotech sector with the necessary insights to drive informed financial decisions across the drug development lifecycle. By integrating economic considerations into decision-making processes, biotech companies can navigate the complexities of drug development, optimize resource utilization, and ultimately contribute to the advancement of transformative therapies for patients in need.

Navigating Regulatory and Reimbursement Challenges

Navigating the intricate landscape of regulatory approvals and reimbursement mechanisms is a critical aspect of financial decision-making in the biotech sector. Biotech companies must align their financial strategies with the evolving regulatory environment and reimbursement dynamics to maximize the market potential of their products.

Regulatory Compliance and Financial Implications

Regulatory compliance entails adherence to stringent guidelines set forth by health authorities to ensure the safety, efficacy, and quality of biotech products. Finance professionals collaborate closely with regulatory teams to anticipate the financial implications of compliance requirements and the potential impact on product development timelines and costs. By integrating regulatory considerations into financial planning, biotech firms can mitigate the risks associated with non-compliance and streamline the pathway to market authorization.

Reimbursement Strategy and Financial Sustainability

Developing a comprehensive reimbursement strategy is imperative for ensuring the financial sustainability of innovative therapies. Finance professionals work in tandem with market access and commercial teams to assess the reimbursement landscape, analyze payer dynamics, and devise pricing models that align with the value proposition of the products. By proactively addressing reimbursement challenges and optimizing pricing structures, biotech companies can enhance the economic viability of their therapies and secure sustainable market access.

Financial Risk Mitigation and Contingency Planning

In the dynamic landscape of healthcare regulations and reimbursement policies, financial risk mitigation and contingency planning are indispensable. Finance experts engage in scenario analysis and risk modeling to anticipate potential regulatory hurdles or reimbursement fluctuations and develop proactive strategies to mitigate their financial impact. By incorporating risk mitigation measures into financial planning, biotech companies can enhance their resilience in the face of regulatory uncertainties and market dynamics.

Leveraging Real-World Evidence for Financial Decision-Making

The utilization of real-world evidence and health outcomes data holds significant implications for financial decision-making in the biotech sector. Finance professionals collaborate with cross-functional teams to leverage RWE in assessing the economic value and clinical outcomes of therapies, informing pricing and access strategies, and enhancing market differentiation. Harnessing RWE can provide valuable insights into the long-term financial implications of innovative therapies and support evidence-based decision-making in the pursuit of financial and patient-centric objectives.

In essence, the integration of regulatory and reimbursement considerations into financial decision-making is pivotal for the success of biotech companies in navigating the complexities of market access and ensuring the sustainable commercialization of transformative therapies. By aligning financial strategies with regulatory and reimbursement dynamics, finance professionals can bolster the strategic positioning of biotech products and ultimately contribute to improving patient outcomes and advancing healthcare delivery.

Bridging the Gap between Innovation and Financial Expertise

In the dynamic and fast-paced field of biotechnology, the collaboration between biotech innovators and financial expertise is paramount for driving sustainable growth and innovation. Financial professionals play a pivotal role in providing strategic insights and support to biotech companies, enabling them to navigate the intricacies of commercialization, regulatory compliance, and market access.

They help in translating scientific innovations into viable financial models, facilitating investment decisions, securing funding, and managing financial risks.

Facilitating Investment Decisions

Financial professionals in the biotech sector facilitate investment decisions by conducting comprehensive analyses of potential opportunities and risks. They assess the financial feasibility of new projects, considering factors such as market demand, competitive landscape, and regulatory considerations. By providing detailed financial models and projections, they enable biotech companies to make informed decisions about resource allocation and investment prioritization.

Securing Funding and Capital Management

Another essential aspect of the collaboration between biotech innovators and financial expertise is securing funding and managing capital effectively. Financial professionals work closely with the executive leadership to identify and pursue funding opportunities, whether through venture capital, public offerings, or strategic partnerships. They also play a crucial role in managing the allocation of capital, ensuring that financial resources are utilized efficiently to support research and development initiatives while maintaining financial sustainability.

Managing Financial Risks

Financial professionals in the biotech sector are responsible for implementing strategies to manage financial risks effectively. They conduct thorough risk assessments, identify potential areas of vulnerability, and develop risk mitigation plans to safeguard the financial health of the organization. By actively monitoring and addressing financial risks, they contribute to the overall stability and resilience of biotech companies in the face of industry challenges and market uncertainties.

Driving Sustainable Growth and Innovation

Ultimately, the collaboration between biotech innovators and financial expertise aims to drive sustainable growth and innovation within the biotechnology sector. Financial professionals play a crucial role in aligning financial strategies with the overarching goals of advancing transformative therapies, contributing to improved patient outcomes, and fostering innovation in healthcare delivery. By providing strategic insights and support, they enable biotech companies to make sound financial decisions that contribute to the long-term success and impact of their innovative initiatives.

In conclusion, the collaboration between biotech innovators and financial expertise is fundamental for achieving sustainable growth, innovation, and success in the dynamic field of biotechnology. Through their strategic insights, financial professionals empower biotech companies to navigate challenges, capitalize on opportunities, and ultimately drive positive change in the healthcare landscape.

Facilitating Strategic Partnerships

Connecting biotech innovators with financial expertise involves facilitating strategic partnerships between finance professionals and biotech research and development teams. By fostering collaboration and mutual understanding, finance experts can gain valuable insights into the scientific and clinical aspects of novel therapies, enabling them to align financial strategies with the specific needs and goals of biotech innovators. This partnership can leverage the financial acumen of professionals to optimize resource allocation, support the development of promising drug candidates, and ultimately drive the successful commercialization of innovative therapies.

Financial Evaluation of Research and Development Projects

Finance professionals play a crucial role in evaluating the financial feasibility and potential return on investment of research and development projects in the biotech sector. By conducting comprehensive financial analyses, they provide valuable insights into the cost of development, anticipated revenue streams, and potential risks associated with innovative projects. Moreover, finance professionals collaborate with R&D teams to establish key performance indicators and financial milestones, enabling the continuous evaluation of project progress and financial outcomes.

Optimizing Resource Allocation

Effective resource allocation is paramount in the realm of biotech research and development, and finance professionals contribute significantly to this endeavor. By leveraging financial expertise, they support R&D teams in allocating resources efficiently, prioritizing projects based on their financial potential, and ensuring that investments align with the organization's strategic objectives. This optimization of resource allocation enhances the likelihood of successful project outcomes and maximizes the impact of financial investments in innovative therapies.

Supporting Market Access Strategies

Finance professionals also play a pivotal role in supporting market access strategies for newly developed therapies. By collaborating with market access and commercial teams, they contribute to the development of pricing and reimbursement models that align with the therapeutic value and market dynamics. This strategic alignment of financial considerations with market access facilitates the successful commercialization of innovative therapies and enhances their accessibility to patients in need.

In essence, the integration of financial expertise into the evaluation of research and development projects in the biotech sector is essential for optimizing resource allocation, supporting market access strategies, and ultimately driving the successful commercialization of transformative therapies.

Leveraging Data Analytics for Financial Insights

Data analytics has emerged as a powerful tool for finance professionals in the biotech sector, offering valuable insights into financial trends, resource optimization, and risk management. By harnessing advanced analytics techniques, finance professionals can gain a comprehensive understanding of financial performance, resource utilization, and market dynamics, enabling data-driven decision-making for sustainable growth and innovation.

Financial Performance Analysis

Data analytics enables finance professionals to conduct in-depth analyses of financial performance, revenue forecasting, and cost optimization. By leveraging historical financial data and market trends, they can identify opportunities for revenue enhancement, cost-saving initiatives, and efficient resource allocation, thereby contributing to the financial sustainability and growth of biotech companies.

Resource Utilization and Efficiency

Through data analytics, finance professionals can gain visibility into the utilization of financial resources across various projects and initiatives. This insight allows them to identify areas of inefficiency, optimize resource allocation, and align investments with high-potential opportunities, ultimately maximizing the impact of financial resources on research and development endeavors.

Risk Management and Mitigation

Data analytics also plays a crucial role in risk management and mitigation within the biotech sector. Finance professionals utilize predictive analytics and risk modeling to anticipate potential financial risks, develop proactive mitigation strategies, and enhance the overall resilience of the organization in the face of market uncertainties and regulatory challenges.

By harnessing the power of data analytics, finance professionals in the biotech sector can drive informed decision-making, optimize resource allocation, and effectively manage financial risks, ultimately contributing to the sustainable growth and success of innovative initiatives.

In conclusion, the strategic integration of data analytics into financial decision-making empowers finance professionals to leverage valuable insights, optimize resource allocation, and effectively manage financial risks, thereby supporting the long-term success and impact of biotech companies in driving innovation and transformative therapies.

The Business of Healthcare: Navigating Financial Challenges

The business of healthcare presents unique financial challenges that demand strategic navigation and innovative solutions. From rising costs to evolving regulatory landscapes, finance professionals play a pivotal role in addressing these challenges and fostering the financial sustainability of biotech companies in the healthcare sector.

Cost Containment and ROI Maximization

One of the primary financial challenges in the business of healthcare is cost containment while maximizing return on investment. Finance professionals work closely with biotech companies to develop comprehensive cost management strategies that streamline operational expenses without compromising the quality of research and development initiatives. They also focus on maximizing the ROI of innovative therapies through strategic financial planning and evaluation.

Regulatory Compliance and Financial Governance

Navigating regulatory compliance and ensuring robust financial governance are critical priorities for biotech companies operating in the healthcare industry. Finance professionals collaborate with legal and compliance teams to establish financial frameworks that align with regulatory requirements, mitigate compliance risks, and uphold transparent financial practices.

Capital Structure and Funding Strategies

The capital structure and funding strategies of biotech companies significantly impact their ability to pursue research and development endeavors. Finance professionals analyze and optimize capital structure, exploring funding avenues such as venture capital, public offerings, and strategic partnerships to secure the necessary resources for innovative initiatives while balancing financial sustainability and growth objectives.

Financial Sustainability Amid Market Volatility

Market volatility and economic uncertainties pose challenges to the financial sustainability of biotech companies. Finance professionals leverage their expertise to navigate market fluctuations, assess investment risks, and devise strategies that fortify the financial resilience of organizations, ensuring their ability to sustain long-term growth despite dynamic market conditions.

In essence, the role of finance professionals is indispensable in navigating the intricate financial landscape of the healthcare sector, ensuring the sustainable growth and success of biotech companies in driving transformative therapies and innovation.

Embracing Technological Advancements for Financial Excellence

In addition to navigating financial challenges, finance professionals in the healthcare sector are increasingly embracing technological advancements to achieve financial excellence. The integration of innovative technologies offers opportunities to streamline financial operations, enhance data security, and drive strategic decision-making.

Automation of Financial Processes

The adoption of advanced financial software and automation tools enables finance professionals to streamline repetitive tasks such as invoicing, budgeting, and financial reporting. This not only reduces manual errors but also frees up valuable time for strategic financial analysis and decision-making, ultimately contributing to operational efficiency and cost savings.

Enhanced Data Security and Privacy

As the healthcare industry grapples with the growing importance of data security and privacy, finance professionals are leveraging technological solutions to safeguard financial information and ensure compliance with stringent regulatory requirements. By implementing robust encryption, access controls, and secure data storage systems, they mitigate the risk of data breaches and uphold the confidentiality of sensitive financial data.

Utilization of Predictive Analytics and AI

Advancements in predictive analytics and artificial intelligence empower finance professionals to forecast financial trends, identify patterns, and make data-driven projections for resource allocation and financial planning. By harnessing the power of AI-driven insights, they can optimize investment decisions, minimize financial risks, and capitalize on emerging opportunities in the healthcare landscape.

In summary, the integration of technological advancements into financial practices equips finance professionals with the tools to enhance operational efficiency, strengthen data security, and harness predictive analytics for informed decision-making, thereby driving financial excellence in the dynamic healthcare sector.

Driving Biomedical Innovation Through Financial Acumen

The innovative and dynamic landscape of biomedical innovation demands a strategic and forward-thinking approach to financial acumen. Finance professionals in the healthcare and biotech sector play a pivotal role in driving the success and impact of transformative biomedical initiatives through their astute financial management and decision-making.

Investment in Research and Development

A key aspect of financial acumen in the biomedical field involves the strategic allocation of resources towards research and development endeavors. Finance professionals collaborate with research and scientific teams to assess the financial viability of innovative projects, optimize funding allocation, and ensure that investment decisions align with the organization's long-term biomedical goals. By effectively managing financial resources, they support the advancement of groundbreaking therapies and medical innovations.

Strategic Partnerships and Collaborations

Finance professionals leverage their acumen to identify strategic partnerships and collaborations that facilitate the exchange of resources, expertise, and funding to drive biomedical innovation.

Through careful financial analysis and due diligence, they evaluate potential partners, negotiate favorable terms, and structure financial agreements that enable the organization to expand its capabilities and accelerate the development of life-changing medical solutions.

Commercialization and Market Penetration

The successful commercialization of biomedical innovations is a critical aspect of financial acumen in the healthcare sector. Finance professionals work in tandem with commercialization teams to develop comprehensive financial strategies that support product launch, market penetration, and revenue generation. By assessing market dynamics, pricing strategies, and return on investment, they contribute to the successful introduction of transformative biomedical solutions to the healthcare market.

Long-Term Financial Sustainability

Ensuring the long-term financial sustainability of biomedical initiatives is a fundamental objective of finance professionals. By conducting thorough financial modeling, scenario analysis, and strategic planning, they pave the way for sustained funding, growth, and profitability within the biomedical sphere. Their proactive approach to financial sustainability supports the continuity of impactful research and development efforts over time.

In essence, the application of financial acumen within the biomedical landscape is essential for driving innovation, fostering strategic partnerships, and ensuring the sustainable commercialization of life-changing biomedical solutions. The collaborative efforts of finance professionals contribute significantly to the advancement of healthcare and the improvement of patient outcomes through transformative biomedical innovation.

Ensuring Regulatory Compliance and Ethical Financial Practices

In addition to driving innovation and strategic decision-making, finance professionals prioritize regulatory compliance and ethical financial practices within the biomedical landscape. Adhering to stringent regulations and ethical guidelines is integral to maintaining the integrity of financial processes and upholding the trust and confidence of stakeholders, including investors, regulatory authorities, and the public.

Compliance with Industry-Specific Regulations

Finance professionals collaborate closely with legal and regulatory affairs teams to ensure adherence to industry-specific regulations and standards governing financial practices within the biomedical sector. This includes compliance with guidelines related to research funding, clinical trials, intellectual property rights, and financial reporting, thereby mitigating the risk of non-compliance and reputational damage.

Transparency and Accountability

Promoting transparency and accountability in financial operations is a cornerstone of ethical financial practices. Finance professionals implement robust internal controls, financial reporting frameworks, and governance structures to uphold transparency and accountability in all financial transactions, fostering trust among stakeholders and reinforcing the organization's commitment to ethical conduct.

Ethical Allocation of Funding and Resources

Ethical considerations permeate the allocation of funding and resources within the biomedical landscape. Finance professionals prioritize the ethical utilization of financial resources, ensuring that investments align with the organization's values, ethical standards, and the broader societal impact of biomedical innovations. This includes conducting due diligence on funding recipients and evaluating the ethical implications of financial decisions.

In conclusion, the role of finance professionals in the biomedical landscape extends beyond financial acumen and strategic decision-making. Their dedication to regulatory compliance, ethical financial practices, and transparency reinforces the integrity of financial operations, ultimately contributing to the sustainable advancement of transformative biomedical innovations and the improvement of healthcare outcomes.

Embracing Technology for Regulatory Compliance

In today's rapidly evolving regulatory landscape, finance professionals in the biomedical sector are leveraging technological tools and solutions to streamline compliance processes and ensure adherence to stringent regulations. Advanced software applications and digital platforms enable automated monitoring of regulatory changes, real-time risk assessment, and enhanced documentation and reporting capabilities. By embracing technology, finance professionals can efficiently navigate complex regulatory requirements while maintaining the highest standards of compliance.

Data Security and Privacy Measures

As the biomedical sector increasingly relies on data-driven insights and technological advancements, finance professionals are at the forefront of safeguarding sensitive financial and patient data. Implementing robust data security protocols, encryption methods, and privacy measures is crucial to protecting confidential information and mitigating the risk of data breaches. By adhering to industry best practices and regulatory guidelines, they uphold the integrity and confidentiality of financial information within the biomedical landscape.

Predictive Analytics for Compliance

The integration of predictive analytics into compliance efforts empowers finance professionals to anticipate regulatory trends, identify potential areas of non-compliance, and proactively address compliance gaps. By analyzing historical data and leveraging predictive modeling, they can enhance risk management strategies, mitigate compliance-related challenges, and adapt their approach to evolving regulatory frameworks, thereby promoting proactive and forward-thinking compliance practices.

Continuous Education and Training

To stay abreast of ever-changing regulatory requirements and ethical standards, finance professionals prioritize continuous education and training in areas such as compliance, ethics, and data privacy. By fostering a culture of continuous learning and professional development, they enhance their expertise in navigating complex regulatory landscapes, understanding emerging compliance obligations, and upholding ethical financial practices, thus contributing to a culture of compliance excellence within the biomedical sector.

In essence, the convergence of technology, data security, predictive analytics, and continuous education amplifies the role of finance professionals in ensuring robust regulatory compliance and ethical financial practices within the dynamic biomedical landscape. By embracing these advancements, they are instrumental in upholding the highest standards of integrity, transparency, and ethical conduct, thereby safeguarding the financial and ethical integrity of biomedical innovation.

1. INTRODUCTION TO HEALTHCARE FINANCE

Experts in healthcare, bioengineering and the life sciences are not often well informed about the fundamentals of commerce and finance. At the same time, financial decision makers may not be knowledgeable about the details of biomedical R&D processes. This book was written to be a bridge between these two sectors – to shine light on the role played by the finance and business world in biotech health care delivery and drug development.

Readers include any stakeholder interested in the crossroads between biotech in healthcare and finance. Across all educational levels, biotechnology decision makers, students and professionals in the biomedical industry, pharmaceutical company executives, investors, portfolio managers, investment bankers, regulators, policymakers, lawyers, patient advocates, patients, philanthropists, trade associations, and so and, are sure to find the contents of this book useful.

The first part of this introductory chapter will look at definitions in biomedicine and, how biomedicine, biotechnology and biopharmaceuticals relate to one another. Next, we'll point out some impactful innovations in the industry, simplify a few basic terminologies in finance, summarize the drug development process, look into a few financial considerations, and then branch out into the book. The majority of the book will then expand on the world of finance and apply them directly to biotechnology (in health care).

Definitions in Biomedicine

To bring about an easier understanding of how they are influenced by the finance sector, in this book, biomedicine, biotechnology (in health care) and biopharmaceuticals will be addressed as one industry. The terms biomedicine, biomed, biotechnology, biotech, biopharmaceuticals and biopharma will therefore be used interchangeably. But firstly, let's briefly define their specifics and see how the three sectors relate to one another.

Biomedicine (biomed) applies molecular biology and physiology to clinical medicine and practice. Biomedical scientists and clinicians focus on medicine and healthcare research to study a variety of conditions in the human body to prevent and treat diseases. Their research contributes to the development of new drugs, treatments and therapies.

Biotechnology (biotech) merges the fields of life science, physical science and engineering to genetically manipulate living organisms to produce new products and processes. It can be applied to healthcare and the broader pharmaceutical industry to develop medications and drugs to treat patients. In fact, biotechnology is a major source of innovation in the pharmaceutical industry.

Pharmaceutical drugs refer to chemically synthesized agents administered to a patient. These are traditional pharmaceutical drugs and medicines. However, biopharmaceuticals (biopharma) refer to the pharmaceutical drug products derived from biotech sources i.e., derived from living organisms and not chemicals. It can therefore be said that biomed and biotech combine in biopharma to develop innovative life-saving drugs, procedures and solutions (Figure 1).

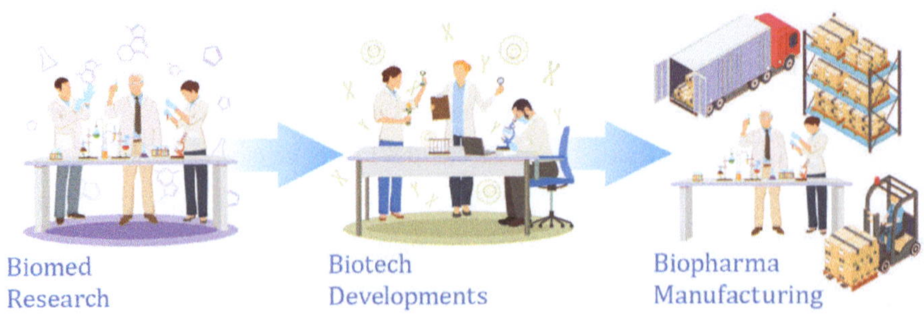

Figure 1: The flow from biomedicine to biotechnology and biopharmaceuticals

On the business and operations side, there is a difference between biotechnology and pharmaceutical companies. On one hand, biotech companies focus on a limited collection of biopharmaceutical drugs in a specific therapy or treatment. They neither develop nor offer an extensive diverse range of drugs and therapies. Venture capital firms may also be their source of business-side operational support. On the other hand, pharma companies do offer a wide variety of different drugs and therapies. Alongside drug discovery and development, they also manage their own marketing and distribution operations.

Bio-Revolution is a newly coined term used to refer to the use of genetic engineering tools to recode biology to treat diseases. The four aspects of bio revolution are biosystems, biomachines, biomolecules and biocomputing.

Biosystems study the complex interactions among cells and larger biological systems. This includes understanding how cells communicate, function, and respond to various stimuli within an organism. An example of innovation in this area is stem-cell technology, which explores the potential of stem cells to regenerate damaged tissues or grow new tissues for therapeutic purposes. Stem-cell research has profound implications for treating a wide range of diseases and injuries, from neurodegenerative disorders to spinal cord injuries.

Biomachines involve integrating machines with the nervous systems of living organisms. This field is at the intersection of biology and engineering, encompassing areas like bio-robotics and neural interfaces. One application is in the development of prosthetics that are controlled by the nervous system, allowing for more natural and intuitive movement for amputees or individuals with paralysis.

Biomolecules, such as DNA, RNA, and proteins, are the building blocks of life. This area of biotechnology explores the role and manipulation of these molecules within cells. Gene therapy for monogenic diseases is a prime example. By targeting specific genes within cells, gene therapy aims to treat or prevent disease. This can involve replacing a mutated gene causing disease with a healthy copy, inactivating a mutated gene, or introducing a new gene to help fight a disease.

Biocomputing involves the use of biological materials to store, retrieve, and process data. This can include using networks of cells or molecules, like DNA, for computational purposes. DNA computing, for example, utilizes the storage capacity of DNA to hold vast amounts of data. This approach is being explored for its potential to surpass traditional computing in handling large datasets, with applications in data storage, bioinformatics, and complex problem-solving.

Each of these areas represents a frontier in biotechnology, offering transformative possibilities for advancing human health, understanding biological systems, and developing new technologies. The interplay of these fields also underscores the interdisciplinary nature of modern scientific research, where biology, technology, and computation converge to create innovative solutions.

Preventive therapy refers to treatments aimed at preventing infections in an uninfected person or treatment aimed at preventing disease spread in an infected person. Curative therapy, in turn, refers to treatments aimed at curing a patient's disease, illness or condition.

Specialty drugs are pharmaceutical agents intended for use against rare medical conditions that affect fewer than 200,000 people. They treat certain cancers, multiple sclerosis, narcolepsy, and other care-intensive conditions. Specialty drugs are made from living organisms and target a specific part of disease processes. On the other hand, non-specialty drugs treat both chronic and acute diseases that affect larger populations and are chemically synthesized.

Specialty drugs are defined by special requirements regarding their storage or handling, how they are administered and how much they cost. Although they were originally designed to treat only rare diseases, as time has progressed, they were developed to treat common chronic diseases.

Orphan drugs are a subclass of specialty drugs. An orphan drug may be a specialty drug, not all specialty drugs have orphan designation. Orphan drugs treat diseases so rare that sponsors are reluctant to develop them under usual marketing conditions. Medicines approved for an orphan indication benefit from incentives such as exclusivity and protection from competition once on the market.

Biologics are another subclass of specialty drugs used to treat patients with severe and life-threatening conditions. They structurally mimic compounds found in the body. More than 60% of orphan drugs are biologics. Biosimilars are a type of biologic that closely mimics an existing, approved biologic (referred to as the reference product). Biosimilars must be clinically similar to the reference product in terms of safety, purity, and potency. They play an essential role in increasing patient access and reducing the cost of important therapies.

Immunotherapy products include biologically derived biologics and chemically derived drugs developed for cancer treatment. They use substances made by the body or in a laboratory to boost the immune system and help the body find and destroy cancer cells.

Monoclonal antibodies (MABs) are a drug class of man-made proteins engineered to serve as substitute antibodies in the human immune system. Gene therapy is a treatment option that modifies a person's genes to treat or cure a disease. Cell therapy (CT) is a treatment option that transplants human cells to replace or repair damaged tissue and/or cells.

A newly developed and approved drug is initially sold under a brand name or trademark. During this time, the drug is protected by a patent which means that only the patent holder is authorized to manufacture, market, sell and make a profit from the drug.

Patents are property rights granted at any time during the drug development process. Drug patents can take 8 to 10 years to be issued and the duration of a patent life is set by statute. Time during a drug's patent life is often lost during the years preceding the drug's approval at the end of the drug development process.

This leaves a reduced amount of time for drug developers and manufacturers to profit from their product. Conversely, the longer a patent on a specified drug, the longer the time a manufacturer will have to profit from market exclusivity. Drug patent extensions can be made to maximize profits. For example, the FDA offers a 7-year extension of patent life for specialty drugs.

A generic drug is a copy of an already marketed brand-name drug. It is developed and manufactured by a company other than the original developer. Generic drugs can only be manufactured and sold by other companies after the patent belonging to the brand name developer has expired.

Generic drugs have an identical clinical composition (in active ingredients, conditions of use, dosage forms and administration methods) with its brand-name drug. It is typically less costly for a generic drug to gain approval than it is for a brand name drug. This encourages competition and results in more competitive drug prices.

An authorized generic drug is an approved brand name drug marketed without the brand name on its label. It is either sold marketed by the brand name drug company, or another company with the brand company's permission. A company may choose to sell their own authorized generic drug at a lower cost than their brand name drug.

Exclusivity is granted on a drug upon approval and refers to delays and prohibitions on the acceptance or approval of any other drug with the same active ingredient. The holder of a new drug application (NDA) or an abbreviated new drug application (ANDA) is eligible for exclusivity if statutory requirements are met. Exclusivity promotes a balance between new drug innovation and greater public access to drugs that result from generic drug competition.

Patents and exclusivity apply to drugs in different ways. For example, patents can be issued or expire at any time regardless of the drug's approval status. Furthermore, the period before exclusivity and/or patent expiration may or may not run simultaneously. Some drugs have both patent and exclusivity protection while others have just one or neither. An Orphan Drug Exclusivity (ODE) period is for 7 years.

Biological applications contribute massively to the improvements in medicine currently experienced globally. Let's look at a few recent breakthroughs.

Biotech Breakthroughs in the Healthcare Sector

Developments in computing, data analytics, machine learning, artificial intelligence (AI), and biological engineering encourages the accelerated breakthroughs seen in biomedicine, biotechnology and biopharmaceuticals. Millions of patients around the world are being treated using newly developed therapies from this industry. Millions more await the next set of innovative biomedical breakthroughs. The industry is at a turning point, continually witnessing new ways of drug and treatment discovery.

One such breakthrough occurred in 2013 when the U.S. Food and Drug Administration approved the first visual prosthesis (also known as a bionic eye). It was called the Argus II Retinal Prosthesis System and made by Second Sight Medical Products. The Argus II replaced photoreceptors in the eye to create artificial vision. It did not allow blind patients to see clearly, but helped them perceive shapes and movement.

The device combined a small video camera and a computer on a pair of eyeglasses and an artificial retina that was implanted inside the eye. The images were wirelessly transmitted from the implant into the artificial retina. However, Second Sight discontinued their retinal implants in 2019.

Another biotech breakthrough is that biological medicine (biologics) grown in live cells now lead to successful treatments. They include antibody-based drugs used to treat diseases such as cancer, rheumatoid arthritis and psoriasis. They also include recombinant proteins that aid diseases such as diabetes or hereditary enzyme-deficiency disorders. 22% of the drugs approved by the US Food and Drug Administration, between 2008 and 2017, were biologics.

Biologics are powerful drugs that derail or stop inflammation that damage joints and organs in inflammatory diseases like arthritis. All drugs have side effects, but for many arthritis patients taking a biologic, the benefits exceed the risks. However, the main challenge is that such drugs are one of the most expensive in the market.

A third noteworthy biotech breakthrough is that T-cells are now being used to fight cancer into long-term remission. This is an immunotherapy treatment made from a patient's own immune system. The CAR-T cells are designed to treat leukemia by harnessing the body's own immune system to target the cancer.

According to a 2018 research paper written by the UK's Royal College of Physicians, over 80% of relapsed Adult B-acute lymphoblastic leukemia (B-ALL) patients with a 7-month life expectancy achieve complete remission following CAR-T cell therapy. This results in 50–86% of patients being alive 1 year later.

In the same way, B-cell lymphoma patients with chemo-refractory disease are given a median life expectancy of 6 months. More than 50% of the patients achieve complete remission following CAR-T cell therapy. This results in over 50% of them being alive 18 months later (although relapses do occur).

By 2022, a 10-year follow up was done on of the first patients treated with CAR-T cells. The follow up concluded that CAR-T cells can actually cure patients with leukemia. The CAR-T cells remained detectable more than ten years after infusion, with sustained remission in infused patients.

A fourth breakthrough is that correct genes are now being introduced into viruses that incorporate the gene into the cells relevant for tissue formation, restoring bodily functions. In 2014, two children diagnosed with a rare genetic retinal disease called Leber Congenital Amaurosis enrolled together in an experimental clinical trial called gene therapy.

By 2021, one of the patients, grown to become a teenager, said of the results of the procedure, "I was supposed to lose all of my sight by age 25 or 30, but the gene therapy that my sister and I got put the gene that we were missing into our eyes. It's helped us see better, but not perfectly. I do not know how it's going to continue for the rest of my life, but so far my vision has been pretty stable since then."

It is evident that the current revolution of healthcare through biotech is quite remarkable. Despite this, financial considerations pose a great challenge in the industry. Before looking into these, let's explain key terms in finance that will be used throughout the book.

Simplifying Financial Terminologies

An asset is anything tangible or intangible that has positive monetary value and can be converted into cash. To invest is to buy assets for the purpose of earning a profit above the initial money spent. Investors come together in financial markets to purchase and exchange these assets (such as stocks or bonds).

A portfolio is one's complete collection of investments. These can be spread across numerous types of investments. Investment types are known as asset classes and are grouped according to the characteristics they exhibit. The most common types of asset classes are cash, equity, fixed-income securities, real estate, commodities, derivatives and currencies.

Diversification refers to spreading investments across assets to reduce overall risk. This gives a literal definition to not putting all of one's eggs in one basket (or type of basket). Then all eggs are not lost at once in case something happens to one of the baskets. Diversification helps to determine what percentage of a portfolio is dedicated to each type of investment. A portfolio holding assets spread out to represent different companies, industries, and is located in different geographical regions, is better built to withstand market turbulence.

Risk refers to the likelihood of loss on an asset, loan, or investment. Risk tolerance is an investor's willingness to accept investment losses in exchange for the potential of higher returns. Investors with longer term goals, for example those spanning over decades, are more equipped to financially (and emotionally) deal with the highs and lows of market volatility. This means they are more equipped to take more risks because there is more time to rebound from market losses.

Possibility of default or *risk of default* is probability that a borrower will be unable to make their scheduled repayments on their debt over a specific period. A higher risk of default means there is a higher chance that a debtor will be unable to pay their scheduled debt payment in a given time period. A lower risk of default means the opposite – there is a higher chance that a debtor will be able to pay their scheduled debt payment in a given time period. These are often confirmed through credit ratings and are an important risk profile tool.

The longer the valley of death, the higher the possibility of default. This is risky to invest in and leaves plenty of room for uncertainty. Will the drug pass through each stage? Will it do so in a timely manner? What is the probability of success at each stage? How much will each stage cost? What are the annual costs?

Financial instruments are contracts that create financial assets for one party and create financial liabilities or equity instruments for another party. They are monetary contracts between parties that can be created, traded, modified, or settled. They can also represent ownership, a loan, or a foreign currency.

Securities are financial instruments. They are any evidence of ownership or debt that has been assigned a value and can be traded. They include debt and equity.

Bonds are an example of a fixed income security. At the basics, they are loan and referred to as debt financing. For a company to issue or sell a bond is for it to borrow and receive a loan. To purchase and hold a bond is to lend money and become a bondholder. These transactions are done by brokers at bond exchanges or through banks.

On the lender side, a bond is an asset and an investment. It's a fixed income because it pays the purchaser fixed regular payments. With bonds, these payments are called coupons. Bonds eventually return the principal (original loan received) at maturity (the expiry date) to the bondholder. It's possible for the initial bondholder to sell the bond to other investors. It does not have to be held until its maturity date.

The market price of a bond depends on the following factors:

- the issuer's credit rating (establishes their coupon/principal payment ability)
- the length of time until expiration (determines for how long interest payments will be made)
- the coupon rate compared to the general interest rate environment at the time

On the lender side, a bond is a source of financing. Corporate bonds do not give ownership in the company. They can be either high-yield or investment-grade.

A high-yield bond has a lower credit rating and offers higher interest rates and higher coupon payments in exchange for a higher risk of default. Investment-grade means the issuer has a higher credit rating and pays lower interest rates and lower coupon payments due to a lower risk of default. Bonds often lose market value when interest rates rise because they are proportional to coupon rates. Borrowers can repurchase their own bonds if interest rates decline, or if their credit improves. They can then reissue new bonds at a lower cost.

Equity is the value of ownership of an investment. It is also referred to as an asset class that includes stocks, mutual funds, hedge funds and private equity.

Stocks are also known as shares. Think of the ownership of a public company as a circle. The circle is divided up to different equal pieces. Each piece is a stock. An investor purchases a stock and owns a part of the company.

A stock therefore represents ownership of a company and a claim on part of the company's earnings. Investors who own stocks in a company are called shareholders or stockholders.

Shareholders receive proportionate income from the shares they own on a routine basis – these are called dividend payments.

Shares are exchanged through a stockbroker. The stock market or equity market is the mechanism for buying and selling shares. A stock exchange is the physical facility where investors trade shares. What a shareholder owns in a company can sometimes be referred to as equity, liable capital or risk capital.

When a private company first sells shares of stock to the public, this process is known as an Initial Public Offering (IPO). In doing so, it is said to go public. A reverse merger is also known as a reverse IPO or a reverse takeover. It is the process by which a publicly listed company is bought by a private company through an asset swap and share issue. A reverse merger is a way for private companies to go public by bypassing the lengthy and complex process of going public.

A Follow-on Public Offering (FPO) is when a company that has completed an IPO issues additional shares to raise additional capital. The process is also known as a secondary offering. Follow-on Financing is when a start-up that has already raised capital privately raises additional capital through another round of funding before going public.

A hedge fund is a partnership of high-income investors that pool a large amount of money together through professional fund managers to earn above-average investment returns. These partnerships have high minimum investment requirements because they pursue a wide range of flexible investment strategies.

Mutual funds allow small or individual investors to pool money together. Each investor receives a proportional share in the gains or losses of the fund. The funds are invested in a professionally managed large collection of investments including, but not limited to, stocks and bonds. Spreading the funds this way is less risky than investing the entire pool in individual stocks or bonds.

A private equity fund is a pooled investment offered by a private equity firm that allows a group of investors to combine their assets to invest in a company or a business. This can be either through buyouts or venture capital. A buyout is when a private equity firm buys and takes control of mature companies with the goal of selling it later at a profit.

A venture capital (VC) is when a private equity firm identifies early-stage start-ups looking to raise cash in exchange for equity in the company. The goal is to invest in companies with high growth potential that can either be sold for a profit at a later date or be taken public through an IPO.

Private investment in public equity (PIPE), refers to when an already-public company offers accredited private institutional investors a certain number of restricted shares at a specified price. These investors include mutual funds or large institutions and they purchase the shares at a below-market price. PIPE transactions enable companies to raise large sums of money quickly.

Private equity valuations are not influenced by the larger market. It is a way to diversify a portfolio and aim for higher returns than at the stock market. Investors in private equity funds are not responsible for managing their newly purchased company.

By nature, biopharma is a risky industry to invest in. Financing is directly related to the risk and uncertainty perceived during the drug development process. Let's summarize the biopharma drug development process and pinpoint how it's related to investor risk and uncertainty.

The Drug Development Process

Drug development refers to drugs and medicines derived from biological sources through biotech developments. However, medical devices, diagnostics and bioinformatics are also developed in biotech for improvements, alongside drugs, to treat patients. Magnetic resonance imaging equipment, dialysis machines and artificial hearts are examples of such medical devices. Blood tests, cancer diagnostics and genetic sequencing are examples of diagnostics. Improvements in these are not exempted from drug development processes.

The drug development process can be divided into three stages (Figure 2).

The first 3-6 years of drug development make up the first stage. These years are for research, discovering the drug candidate and preclinical studies. Preclinical studies may involve in-vivo studies (for example in petri dishes) and in-vitro studies (for example testing in animal models like mice or monkeys).

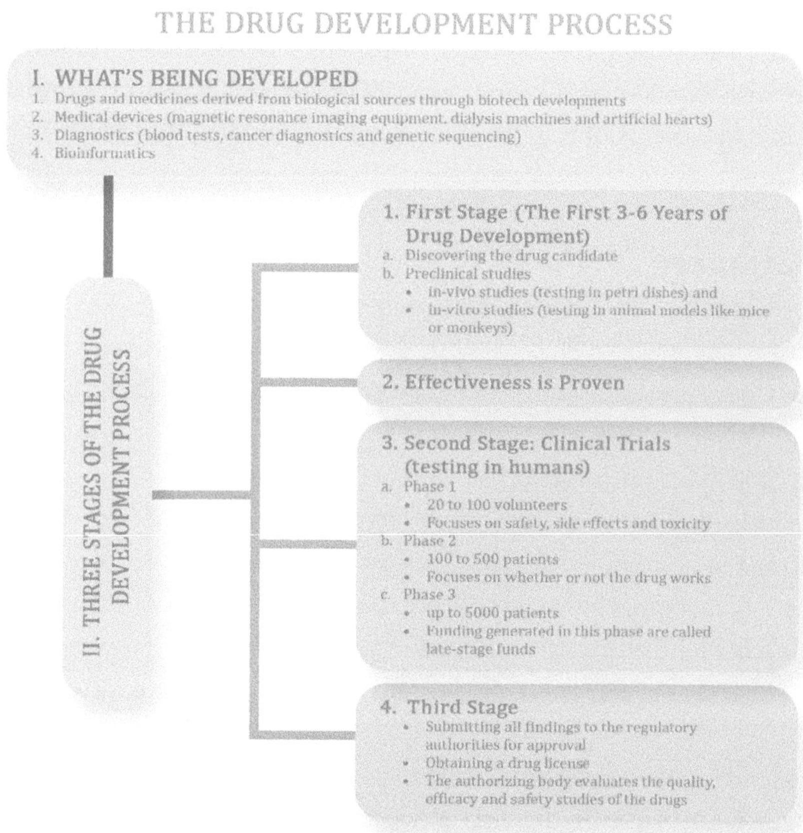

Figure 2: Stages of the Drug Development Process

Once effectiveness is proven, the developed drug can reach the second stage: clinical trials. This is when testing is done in humans. Phase 1 of clinical trials involve 20 to 100 volunteers and focuses on safety, side effects and toxicity. Phase 2 involves 100 to 500 patients and focuses on whether or not the drug works. Phase 3 involves a broader population of volunteers of up to 5000 patients. Funding generated in this phase are often referred to as late-stage funds.

The third stage of the of drug development process involves submitting all findings to the regulatory authorities for approval and obtaining a drug license. The authorizing body evaluates the quality, efficacy and safety studies of the drugs. The authorizing body in the US is the Food and Drug Administration (FDA). In the EU, it is the European Medicines Agency (EMA). In the UK, it is the Medicines and Healthcare products Regulatory Agency (MHRA). In Türkiye, it is the Turkish Pharmaceuticals and Medical Devices Agency (TİTCK).

Once the drug is approved and licensed by the relevant authority, sales permits are granted. This is when the drug can be put on the market and manufactured for mass medical consumption. After this, studies, changes and updates regarding the drug are expected to be carried out by the drug developer throughout the life-period of the drug.

The entire process of drug development, from the early stages of the drug discovery to being granted sales permits, can take up to 20 years. This leaves a large time gap until the drug generates revenue. This time gap is known as the *valley of death.*

The time gap affects how long it will take investors to start making money off their investment. A longer time gap also equates to a higher possibility of default and therefore, higher risk and uncertainty. This is how aspects of the drug development process directly influences funding and other financial considerations.

Financial Considerations

The many different processes involved in drug development and delivery from the laboratory to the patient, all have funding as a common denominator. As a result, business models, investor behavior, risk and reward trade-offs, and other financial considerations automatically play a significant role in biomedical innovation.

Funding is promised if biotech developers have promising evidence-backed developments. Conversely, the processes involved in reaching said evidence all require funding to cover the cost of laboratory space, equipment, supplies, patient volunteers, and the salaries of scientists, engineers and clinicians. Funding is also needed to cover the filing fees for regulatory approval as well as the cost of manufacturing, marketing, distributing, and monitoring the approved drug.

Putting these factors into consideration often lead biotech decision makers to propose therapies for funding in accordance with the drug or solution's estimated market value. This is an example of how source of financing can drive the scientific agenda.

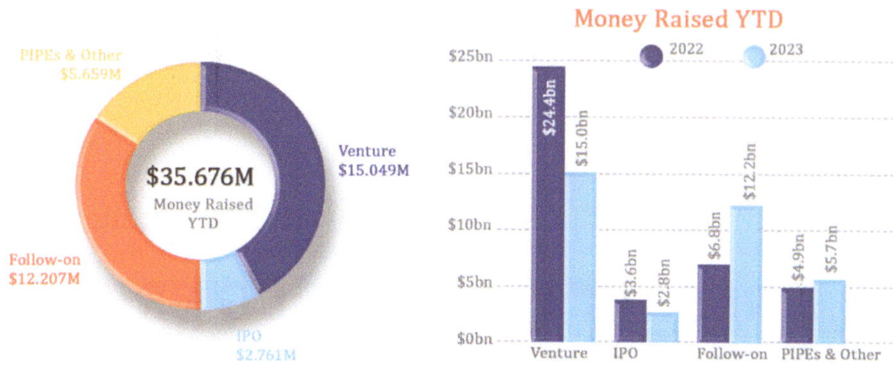

Figure 3: Financing Options for Biotech and Biopharma (Source: BioCentury)

Biotechnology and biopharma companies mostly receive heavy financing from either initial public offerings (IPOs), healthcare-focused venture capitalist funds, PIPEs or FPOs (Figure 3). These sums come up to billions of dollars in annual funding.

In 2021, biotech companies raised a global total of $29 billion in capital through IPOs. By August 2023, the year-to-date funding from IPOs was down to $2 billion – a 93% decrease. The recent trend over the years is that venture capital financing is exceeding all the other financing sources. This suggests that VCs have substantial funding reserves to deploy.

2. PRESENT VALUE, DRUG PRICING MODELS AND INFLATION

One of the preliminary aspects of finance is the study of cash flow. Cash flow is the exchange of money between two parties, and all business decisions are based on this. However, the details of these cash flows can be extremely difficult to understand, particularly if there are any uncertainties about their size or timing.

This chapter explains the vocabulary and proper analytical framework necessary for developing the know-how for sound financial decision-making.

The financial system enables the transfer of money across time and space. This function is especially important for the biopharma industry because drug development can be done across the globe, take over a decade, and necessitate an enormous amount of money per drug.

Money can be transferred in time to the future through investments in financial assets. The easiest example of transferring assets from the present to the future is when funds are deposited into a bank account one day and withdrawn later.

On the other hand, taking out a loan moves money backward from the future to the present. Whereas the lender providing the loan does the opposite. Lending money in the present and receiving repayment with interest, in the future, moves money forward in time.

To understand how resources are transferred across time and space, let's begin by analyzing the time value of money.

Fundamental Challenges of Finance

Finance is the language by which business is conducted. Across all sectors and industries, negotiation involves managing cash flows and managing risk. This is essentially what corporate financial managers and entrepreneurs are responsible for.

There are three aspects of managing cash flows. The first aspect is deciding *what* to buy or *what* to sell. Whether you are running a start-up or overseeing a Fortune 500 corporation, dealing with this is a frequently encountered problem. The second aspect is deciding *when* to buy or sell. The more taxing aspect of managing cash flows is the third aspect which is deciding *how* to pay or *how* to finance.

These are the three vital questions that need to be answered. In biomedicine, deciding on these questions is made difficult because of the complexity of finding out how much size has a real impact on investment decisions. They are also difficult to answer because of the time stretch of the previously explained *valley of death* and because the large investment sums are made in whole. Meaning that a percentage of a drug cannot be invested in. You either do the drug or you do not do the drug.

Both risk and uncertainty also play a role in increasing the complexity of answering these questions in biotech. Meanwhile, over the course of the last few years, these complexities have been increasing. Making the process of managing financial assets in biotech even more complicated.

What is an Asset?

In biomedicine, the definition of an asset takes on an important dimension. In a normal context, assets are physical goods, services and ideas. For example, a patent can also be bought or sold. However, additionally, the flexibility to create and launch new projects is also an asset.

There are ranges of things that are not typically considered an asset that can be taken into account in biomedicine. For example, knowledge, reputation, and opportunities are assets as well. Although the challenge can be how to value them.

From a business perspective, an asset is a series of positive or negative cash flows (CF). Positive cash flows are money coming in. Negative cash flows are money going out.

$$\text{Asset}_t \equiv \{CF_t, CF_{t+1}, CF_{t+2}, \ldots\} \qquad (2.1)$$

Irrespective of what can be considered an asset, the main thing to be knowledgeable about is the series of cash flows that anything referred to as an asset generates. This simplifies any approach to applying analytics and frameworks that expand on it.

Therefore, to valuate an asset is to valuate the asset's series of cash flows. *Series* here refers to the fact that the order of occurrence is important i.e. identifying exactly when the cash flows are coming in and going out and putting them in order.

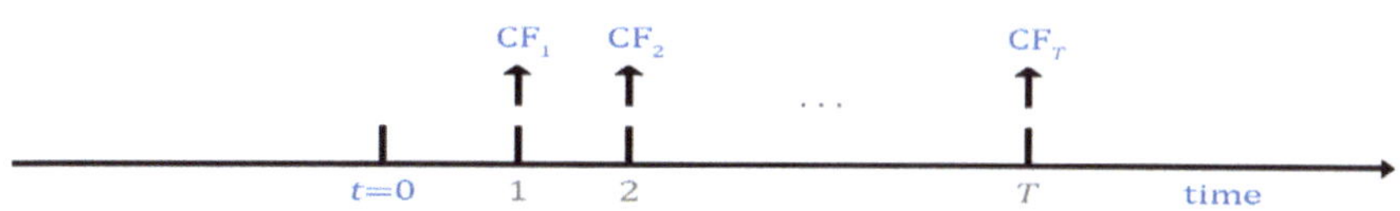

$$\text{Value of Asset}_t \equiv V_t(CF_t, CF_{t+1}, CF_{t+2}, \ldots) \qquad (2.2)$$

To calculate the value of a series of cash flows, the present value operator is constructed.

The Present Value Operator

Cash flows at different dates are not worth the same because without first converting them, past and future cash flows cannot be evaluated together. They can be viewed as different currency because they are worth differently. Each date can be said to have its own exchange rate with another date. The value of a cash flow today is labelled as the *present*. Present can be viewed as the day that the exchange rate converts currencies into.

$$V_0(CF_0, CF_1, \ldots) = CF_0 + \begin{pmatrix} \$0 \\ \$1 \end{pmatrix} \times CF_1 + \begin{pmatrix} \$0 \\ \$2 \end{pmatrix} \times CF_2 + \cdots \quad (2.3)$$

Where:

V_0 is the Net Present Value (NPV)

CF_t can be negative (future costs)

$CF_t < 0$ if there is an initial investment

With *present* being the relevant benchmark, the value of $1 today is the reserve currency of all analyses. For example, what is the exchange rate between $1 tomorrow versus $1 dollar today, $1 next year versus $1 today, $1 three years from now versus $1 today? Calculating all of these various exchange rates gives the present value operator.

The present value operator tries to figure out what a series of cash flows is worth and can therefore be viewed in the same way as a foreign currency translator. It converts the value of a cash flow on one date to the value it would be on another date.

So, the idea behind time value relations and net present value is to use a common currency or equivalent valuation. From here, the value of any series of cash flows can be calculated.

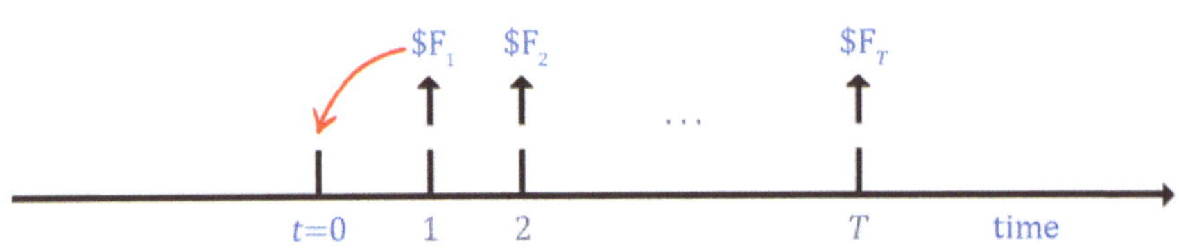

$$V_0(CF_1, CF_2, CF_3, \ldots) = \begin{pmatrix} \$0 \\ \$1 \end{pmatrix} \times CF_1 + \begin{pmatrix} \$0 \\ \$2 \end{pmatrix} \times CF_2 + \cdots \quad (2.4)$$

The expression for the present value operator is a linear function. It multiplies the cash flows by what they convert to and generate an equivalent currency of dollars in the present. Then adds up all the figures. The recommendation is then to keep projects with positive net present value and sell or dispose of projects that have a negative net present value. Let's look at an example.

Suppose the following:

$$\binom{\$0}{\$1} = 0.90\ ,\ \binom{\$0}{\$2} = 0.80$$

What is the net present value of a project requiring a current investment of \$10 million with cashflows of \$5 million in Year 1 and \$7 million in Year 2?

$$V_0(CF_0, CF_1, \ldots) = CF_0 + \binom{\$0}{\$1} \times CF_1 + \binom{\$0}{\$2} \times CF_2 + \cdots (2.3)$$

Where:

Current investment = CF_0 and is a negative figure ∴ $CF_0 = -\$10$

Cash flow in Year 1 = CF_1 = \$5

Cash flow in Year 2 = CF_2 = \$7

$$\text{NPV}_0 = -\$10 + \$5 \times 0.90 + \$7 \times 0.80 = \$0.10$$

The net present value is \$100,000 which is positive. It should therefore be kept because it's positive.

The Time Value of Money

What determines the value of a dollar today versus next year or the value of a dollar today versus two years from now? The answer is a calculation that presumes that \$1 next year will be discounted at a specified rate.

The rate, *r*, is called the interest rate. It is also called the cost of capital, opportunity cost of capital or discount rate.

$$\frac{\$1}{(1+r)} \text{ in Year 0} = \$1 \text{ in Year 1}$$

$$\frac{\$1}{(1+r)^2} \text{ in Year 0} = \$1 \text{ in Year 2}$$

$$\vdots$$

$$\frac{\$1}{(1+r)^T} \text{ in Year 0} = \$1 \text{ in Year T} \qquad (2.5)$$

The idea is to convert currency from one date to another. So, from this year, \$1 next year is going to be worth .

The rate, *r*, discounts the future to the past. Supposing that r is greater than 0, is always less than 1. That is to say, these interest rates are usually positive concludes that people prefer more money now to later.

The next calculation would be that $1 two years from now converted to today assumes it's , then for three years from now, then for four years from now, and so on and so forth.

It also works the other way around. If we want to take $1 and let it grow over time, $1 today will be worth more than $1 next year.

$$\$1 \text{ in Year } 0 = \$1 \times (1+r) \text{ in Year } 1$$

$$\$1 \text{ in Year } 0 = \$1 \times (1+r)^2 \text{ in Year } 2$$

$$\vdots$$

$$\$1 \text{ in Year } 0 = \$1 \times (1+r)^T \text{ in Year T} \qquad (2.6)$$

Where *T* is the number of time periods that we are trying to convert currencies between and assuming the rate is constant, shows the rate of decay experienced with these kinds of processes. What this tells us is that using this framework, in the distant future, $1 now has almost no value.

As seen in Figure 4, a lower rate gives a slower rate of decay, and a higher rate gives a faster rate of decay. On the green curve versus the blue curve, the blue curve discounts the future more quickly than the green curve because it has a higher rate of 12%.

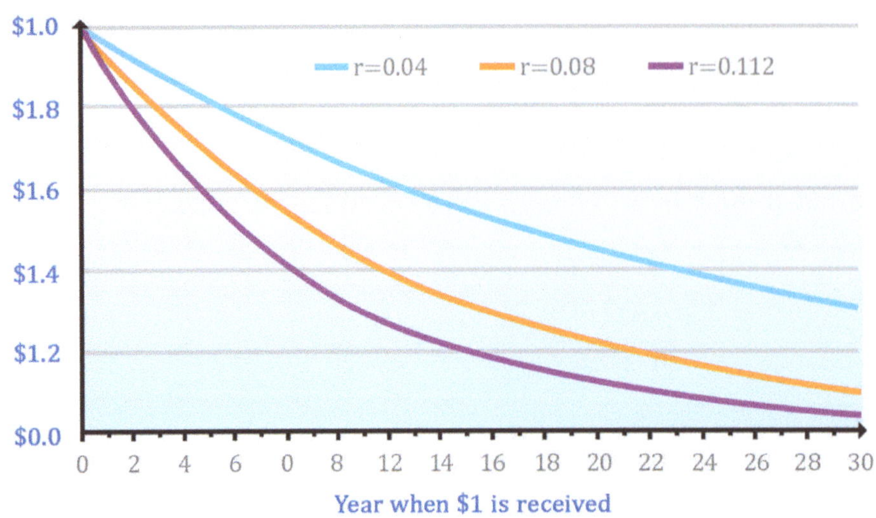

Figure 4: Rate of Decay

This framework is used by the majority of individuals and decision makers to place a value on future assets. Projects are often ranked according to the magnitudes of the net present value. Capital budgeting and corporate finance decisions incorporate cost of capital rates.

However, where the time value of money concludes that the future value of a commodity does not matter from a financial perspective, it might be wise to bring in other considerations to come to a more holistic conclusion. For example, linking the time value of money back to biomedicine, when analyzing a commodity like a vaccine, the numbers do not work if you just look in the present value of cash flows. Other frameworks and arguments that will influence financial decision makers need to be brought up.

The framework for the present value operator and time value of money assumes that the rates are constant over time and markets are perfect. The reality though is that biomedical markets are not perfect. Methods to deal with those frictions will be looked at later in the book.

Next, let's take this framework and see how to valuate special cash flows. The first would be perpetuities and the next would be annuities.

Perpetuities

The Perpetuity

The first formula to look at for valuating special cash flows is the perpetuity formula. Perpetuity pays constant cashflow C forever. Here, a cash flow pays the same amount of money every year, infinitely. Calculating the value is straightforward if the discount rate is positive. The perpetuity formula uses simple algebra as seen below.

$$PV = \frac{C}{(1+r)} + \frac{C}{(1+r)^2} + \frac{C}{(1+r)^3} + \cdots$$

$$(1+r) \times PV = C + \frac{C}{(1+r)} + \frac{C}{(1+r)^2} + \cdots$$

$$r \times PV = C$$

$$PV = \frac{C}{r} \tag{2.7}$$

The present value of a perpetuity that pays, starting next year, C dollars every year out to infinity, has a finite value. Its worth is C divided by the interest rate.

For example, if C is $1,000, meaning $1,000 is paid every year *ad infinitum*, and the discount rate is 10%, then the present value is equal to $10,000.

The formula here consists of the time value of money and how discounting works over time.

The Growing Perpetuity

The second formula to look at for valuating special cash flows is the growing perpetuity. This is where the cash flows increase over time and are paid out infinitely.

The idea behind growing perpetuity is that it starts off next year with cash flow . Then the year after, the cash flow is times 1 plus the growth rate, . Then the year after that, the cash flow is times 1 plus squared and so on and so forth.

$$PV = \frac{C}{(1+r)} + \frac{C(1+g)}{(1+r)^2} + \frac{C(1+g)^2}{(1+r)^3} + \cdots$$

$$\frac{(1+r)}{(1+g)} \times PV = \frac{C}{(1+g)} + \frac{C}{(1+r)} + \frac{C(1+g)}{(1+r)^2} + \cdots$$

$$\left[\frac{(1+r)}{(1+g)} - 1\right] \times PV = \frac{C}{(1+g)}$$

$$PV = \frac{C}{r-g} \quad , \quad r > g \qquad (2.8)$$

Standard arithmetic is then used to calculate the present value. The present value is calculated as seen in Equation (2.8), where is assumed always to be greater than .

The conclusion here is that cash flows increasing over time will be worth more than if they're not increasing. It's possible to quantify how much more they're worth by looking at the and getting a sense of the value of that kind of growth.

For example, a company receives $1,000 every year starting next year, the growth rate is 5%, and the discount rate is 10%. Putting it into the formula, because this is a growing perpetuity, the present value is not $10,000, but $20,000.

$$C = \$1{,}000, \ r = 10\%, \ g = 5\%$$

$$\Longrightarrow$$

$$PV = \$20{,}000$$

It's important to note that with a growing perpetuity, any change in the growth rate, g, can lead to some significant changes in market value because the denominator, , is sensitive to perturbations.

A real-life example of this would be to look at a company expected to be on a growth path and doing progressively better over time. We can look at when Puma Biotechnology was looking to expand into a biopharmaceutical company.

In July 2014 when their breast cancer drug, neratinib, was in clinical trials, the company saw a remarkable increase in the value of their shares. In one day, their stock prices leapt from approximately $60 per share to approximately $230 per share and peaked at $270 per share by September of the same year.

The hike occurred because neratinib's phase III trial for breast cancer had reached its clinical endpoint and phase III had seen particularly promising results. The share prices have however gone back down since then for a number of other reasons.

One reason is that of the side effects of neratinib is severe diarrhea. The company unfortunately found that it was not easy to control the diarrhea with other antidiuretics. This is an example of how the quickly the market moves happen because small changes in growth rate, .

Aside from clinical trial progress, a drug approval is another change in growth rate, g, that causes the market to update a company's value. By July 2017, neratinib was approved for medical use in the United States and was approved for medical use in the European Union by August 2018.

This was a brief illustration of why biopharma companies experience more volatility than companies in other industries. They undergo rapid information shocks through binary events, for instance where drug approval or rejection results are released in the course of a few hours. That's what causes the fluctuating market swings in biopharma.

Annuities

The third formula to look at for valuating special cash flows is the annuity formula. It is used quite frequently in finance. The annuity formula can be used to analyze what one is owed or what one has to pay, for example, the formula to calculate a mortgage, an auto loan or credit card receivables.

$$PV = \frac{C}{(1+r)} + \cdots + \frac{C}{(1+r)^T}$$

$$(1+r) \times PV = C + \frac{C}{(1+r)} + \frac{C}{(1+r)^{T-1}}$$

$$r \times PV = C - \frac{C}{(1+r)^T}$$

$$PV = \frac{C}{r} - \frac{C}{r}\frac{1}{(1+r)^T} \quad (2.9)$$

The idea behind annuities is that there is a finite amount of identical cash flows, generated by buying a perpetuity, holding onto it for T periods, and then selling the perpetuity. At which point the cash flow stops. Engaging in that transaction replicates the cash flow of an annuity.

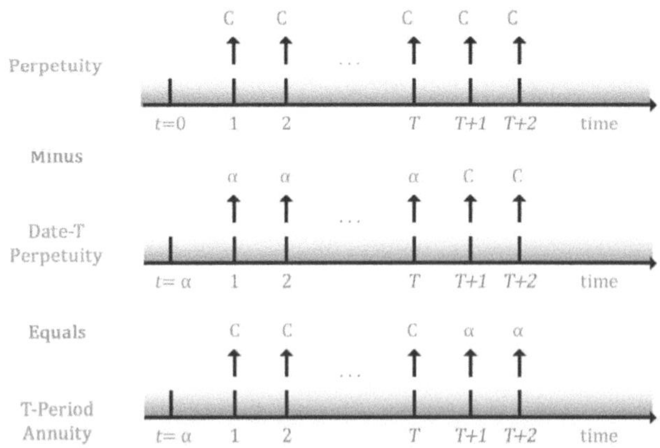

Figure 5: Perpetuity

If the perpetuity were bought today, dollars would be paid for it. Then there's a wait period of T periods to enjoy the cash flows. To sell the annuity would be dollars because it's going out into the forthcoming future. However, since the annuity would be sold after period T, the proceeds are really worth discounted back T periods.

For buying the annuity, is gotten. Meanwhile, is gotten in rebate from selling the annuity and then subtracted from . This is the annuity discount formula. It's the basis of mortgages, of all sorts of consumer loans.

Next, we'll look at how these are applied in equities.

Three Pricing Formulas

The three pricing formulas used regularly for financial analysis are the price of a perpetuity, the price of an annuity, and the price of a growing perpetuity also known as the Gordon Growth Model.

Perpetuity pays constant cashflow C infinitely:

$$PV = \frac{C}{r} \tag{2.7}$$

Annuity pays constant cashflow C for T periods (Equation):

$$PV = \frac{C}{r} - \frac{C}{r}\frac{1}{(1+r)^T} \tag{2.9}$$

Growing perpetuity pays growing cashflow $C(1+g)^T$ infinitely:

$$PV = \frac{C}{r-g} \quad , \quad r > g \tag{2.8}$$

These formulas are the building blocks of all of bond pricing under certainty, if the payments and timing are known. The mode of analysis is however changed if there's any randomness – which is how valuation models for stocks and risky bonds are calculated. In other words, if is constant and known, the formulas are used to valuate bonds. Whereas if is random, the formulas are used to valuate stocks.

Inflation

Pharma spends a huge amount of money on R&D. According to the industry trade group, FHRMA, global pharma R&D spending back in 1980 was $2 billion. As of 2015, $58.8 billion was spent in R&D. As seen in the Figure 4 below, Statista's 2023 estimation of what global pharma R&D spending will be by 2028 is $285 billion. The overall trend seen is that pharma spends more on research in one year than it would the year before.

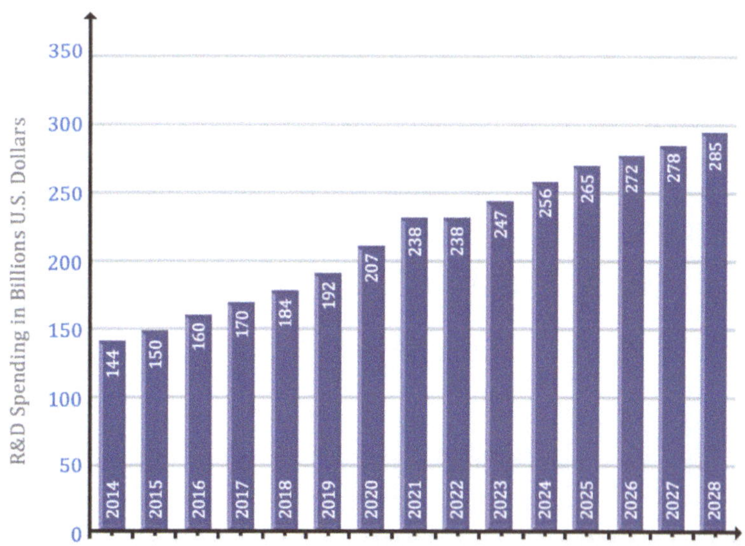

Figure 6: Estimation of Global pharma R&D spending (Source: Statista)

Why are biomedical expenses going up? Perhaps because research and developments should increase in order to keep up with all the biotech breakthroughs occurring? Which in turn leads to the difference in costs? Perhaps.

There is also the concept of inflation to look into. Inflation is an overall increase in prices and decrease in the purchasing value of money. It's the idea that with one legal tender note, you can purchase an item. Whereas after inflation, the same note can no longer afford the same item.

Some terms used in inflation calculations are consumption basket, nominal and real. A consumption basket represents consumer spending and is used to track changes in the prices of goods and services over time. A nominal value is the current monetary value and is not adjusted for the effects of inflation. A real value is nominal value adjusted for inflation.

Inflation is an average across a variety of different consumption goods. It is calculated as the cost of a consumption basket next year divided by the cost of the same consumption basket this year, minus 1. The inflation rate across an entire basket is a weighted average of the inflations of each commodity.

Inflation can be calculated as:

$$i = \frac{C_t + 1}{C_t} - 1 \qquad (2.10)$$

For example, the inflation rate in the United States, is approximately 2%. This 2% inflation rate is a weighted average of housing inflation, transportation inflation, education inflation, medical inflation, and so on.

A negative inflation occurs when prices of a consumption basket go down and purchasing value goes up. This called a deflation. When industries experience a deflation, other factors go into play. For example, improvements in productivity and technological advancements.

Inflation measures how the population is doing relative to the things that they consume. As the consumption baskets change, different measures of how well the population is doing are received.

CPI and BRDPI

The most well-known indicator of inflation is the Consumer Price Index (CPI), which measures the percentage change in the price of the items in the consumption baskets of households. It is used for consumers.

For producers, a producer price index (PPI) is used. In the biomedical industry, the Biomedical RND Price Index (BRDPI) is used. BRDPI was created by the National Institutes of Health (NIH) as a basket what NIH spends on its various different baskets of goods. It was constructed to give law makers a sense of what NIH was doing with its funding and to see if the funds were enough to cover NIH's various different expenses. BRDPI also measures how quickly those expenses were changing and over the years, it has been increasing much faster than the rate of inflation.

For NPV calculations, nominal rates are the actual prevailing interest rate while real rates are the interest rates adjusted for inflation. Nominal cashflows are expressed in actual-dollar cashflows. Real cashflows are expressed in constant purchasing power. Real cashflows are discounted using real interest rates () while nominal cashflows are discounted using nominal interest rates ().

Occasionally, an approximation of the generic basket of commodities and services that the general population consumes is used as the rate of inflation. The real rate is then calculated as the nominal rate minus the inflation rate.

$$(1 + r_{real}) = \frac{(1 + r_{nominal})}{(1 + i)} \approx 1 + r_{nominal} - i$$

$$r_{real} \approx r_{nominal} - i \qquad (2.11)$$

Inflation affects people in different ways. To evaluate if a population is better off or worse off, it's first of all important to decipher what they want. What the population wants is reflected in what they are consuming. Their consumption basket.

For example, if one has $100 and can purchase something essential for $10, they can afford 10 of those items with their $100. If the next day, the item costs $25 (due to inflation), they can only afford 4 of those essential items. They're worse off because more money is needed to put them where they used to be.

The formal measurement of inflation starts off by finding nominal wealth . Nominal wealth is the total of amount of money at the bank while is the price of the basket of goods and services that the population likes to consume. Here, to be better off is to consume goods and services that have gotten much cheaper while to be worse off is to consume goods and services that have gotten more expensive.

$$W_t \equiv \text{Nominal Wealth}$$

$$C_t \equiv \text{Price Index of Consumption Basket}$$

$$= P_{1t}Q_1 + P_{2t}Q_2 + \cdots + P_{nt}Q_n$$

$$\Rightarrow \qquad \tilde{W}_t \equiv W_t/C_t \text{ ('Real Wealth' at date } t\text{)}$$

$$\tilde{W}_{t+1} \equiv W_{t+1}/C_{t+1} \text{ ('Real Wealth' at date } t+1\text{)}$$

$$\tilde{W}_{t+1}/\tilde{W}_t \equiv \text{'Real' Return from } t \text{ to } t+1$$

$$(1 + r_{real}) = \frac{W_{t+1}/W_t}{C_{t+1}/C_t} = \frac{(1 + r_{nominal})}{(1 + i)} \tag{2.12}$$

Haven calculated the price index of what the population likes to consume, the next calculation is to find the real wealth. Real wealth is the real amount of money held in units of things the population likes to buy. Real wealth is then calculated as nominal wealth divided by the cost of the consumption basket.

Change in real wealth depends on two things: investments in nominal wealth as well as the prices of the items in the consumption basket. The rate of return on real wealth is therefore calculated as the rate of return on nominal wealth divided by how quickly the consumption basket is increasing.

Drug Pricing Models

Ultimately, taxpayers pay for drugs and therapies through the government. The government also intervenes in safeguarding reasonably priced access to treatment through partial pharmaceutical price regulation. The scope of these regulations could be in reference to what other countries pay, the prices of existing competitors or through economic evaluation.

In the long run, insurers and health insurance coverage subsidize pharmaceutical expenses and pay for the costs of drugs and therapies. Insurance companies look out over 10 to 30-year horizons and focus on the long run. This helps to minimize financial barriers to healthcare access and usage.

However, the more pressing and proximate financial burden of expensive drugs and therapies falls on patients, families and foundations who find funding through consumer loans and charities. Expensive drugs therefore typically have direct short-term cost implications for the budgets of healthcare payers.

Finance can play a role in reducing the heavy, immediate cost burden of expensive drugs for both public and private healthcare payers. After years of experimenting with different financial arrangements for different product types, the finance industry now proposes a few suggestions for addressing the way medicines are bought and sold globally. These newly proposed drug pricing models aim to provide feasible and long-term solutions that, moving forward, manage the payments for high-cost drugs.

Five novel drug pricing models are the value-based pricing model, the mortgage pricing model, the subscription pricing model, the Indication-specific pricing model and the volume-based pricing model.

The value-based pricing model

Traditionally, biopharma product prices are set based on data and evidence collected during the R&D and clinical trial processes. So, drug pricing has always been contingent on how much was spent to develop the drug.

The proposal behind value-based pricing models, on the other hand, is to place a price on drugs based on how it performs after being administered to real-world patients. This performance can either be based on financial risk or based on health outcomes.

Value-based pricing models also advocates for healthcare payers to enter into pricing agreements directly with the drug manufacturer and eliminate a middle-man.

The suggestion behind the financial risk performance is that payers are reimbursed partially or completely according to financial considerations drawn-up prior to concluding on the pricing contract. The financial consideration could be whether or not the drug is able to treat the patient for a lower price than that of a competitor. Another financial consideration could be whether or not additional hospitalization would be required.

The suggestion behind the health outcomes performance is that payers are reimbursed according to the patient's response to treatment or the extent to which a targeted health outcome is reached. Specialty products in the primary care market segment are the proposed candidates for exploring health outcomes-based contracts. For example, curative therapies for hepatitis C offering refunds for patients who are not cured after using the therapy.

The mortgage pricing model

A high-cost drugs-mortgage pricing model involves amortizing an expensive drug's costs over time. This suggested model for purchasing drugs offers to minimize the burden of paying the total cost of an expensive drug upfront. It allows purchasers to spread the cost of an expensive treatment over time similar to housing mortgage payments.

The idea of spreading out high-cost drugs over a period of time was first conceived in the context of cancer drugs at the turn of the 21st century. Panel discussions held in that time discussed the obstacles related to paying for cure yielding drugs, and reached a unanimous agreement that financing mechanisms should be drawn; that drug purchasing financing mechanisms needed to be revised to eliminate downstream, long-term costs when a cure or a treatment that leads to a cure enters into the market.

One such suggestion was for manufacturers to participate in drug rebate agreements with the state in return for state funded health insurance covering most of the manufacturer's drugs. An example of how this has been implemented can be seen in the United States Medicaid Drug Rebate Program that provides for mandatory rebates on innovator drugs (brand drugs), blood clotting factors, drugs Food and Drug Administration-approved exclusively for pediatric indications, and non-innovator drugs (generic drugs).

The Medicaid Drug Rebate Program maximum rebate is capped at 100% of the Average Manufacturer Price (AMP). The mandatory rebate is calculated as a base rebate plus an inflationary rebate. The base rebate is the greater of 23.1% of the AMP per unit, or the difference between the AMP and the best price offered to most commercial insurers.

The inflationary rebate is calculated as the difference between the AMP on the launch date compared with the current quarter and the United States Consumer Price Index-Urban. Furthermore, if a manufacturer enters into a contract outside of the Medicaid program with a price is below those provided under the Medicaid Drug Rebate Program, a wholesale change in price would need to be performed to ensure that the best price is offered to Medicaid.

Mortgage models can decrease the total upfront payment attached to high-priced drugs and promote better affordability for patients. New immunotherapy products for oncology, orphan drugs targeting rare diseases, or emerging gene and cell therapies, are all good candidates for the mortgage model.

The subscription pricing model

The subscription model proposes an "all you can treat" system for the purchase of drugs and was first initiated in Australia for hepatitis C products. The model allows for unlimited patient access to a certain volume of specific drugs at a fixed recurring price for a set period of time. This model by accelerates patient access and reduces long-term societal costs by removing benefit design restrictions.

States can hit the target product volume by pooling consumers across health care delivery systems, such as Medicaid and correctional settings. Additionally, since treatment populations are prone to unforeseen fluctuations, manufacturers taking subscription models into consideration should clearly define the patient population accessing it.

Subscription models may be useful for curative therapies such as gene and cell therapies. Another suggestion is for it to be applied to insulin manufacturers to offer diabetes patients a monthly flat price for a monthly supply of up to a certain number of boxes of insulin pens.

The indication-specific pricing model

The proposal behind the indication-specific pricing model is to place a price on drugs based on the therapeutic area it is prescribed to treat. Contracts under this model have the payer pay more for a drug when it's highly effective in treating one disease and pay less for treatments where the same drug is less effective.

The indication-specific drug pricing model is used for drugs approved to treat more than one type of disease. For example, monoclonal antibodies are often granted approval for multiple treatment indications including cancer, asthma, macular degeneration, arthritis and Crohn's disease. The ongoing trend is now to use the indication-specific pricing for monoclonal antibodies.

The volume-based pricing model

Volume pricing involves dropping the price per unit of a product as the quantity of the order increases. Volume pricing contracts can be effective in selling more products by having payers purchase a higher quantity due to the competitive prices offered.

Volume-based drug pricing models are less appropriate for high-cost specialty drugs or biologics and are more appropriate for instances where large quantities of drugs are required. Suitable payers for this pricing model include government payers that provide free vaccines to citizens.

3. CAPITAL BUDGETING

This chapter looks closely into valuation to assess the logic behind why some companies are worth $1 billion while other companies are not. Any individual looking into starting companies should be curious about how valuations come about. Insight on patient access is also essential because ultimately, patients need to have access to the amazing therapies developed to help them.

Scientists and biologists have the point of view of the fundamental science regarding the therapies that they develop. However, in order to ensure patient access to them, having the point of view that analyses the plausibility of the financial management aspects is just as important.

Over the course of the book, starting with this chapter, those financial management aspects will be described. This includes the basics about valuation in general as well as in the context of biomedicine. We'll also look into capital budgeting and the capital asset pricing model (CAPM).

Capital budgeting involves decisions pertaining to whether or not to invest in a project while the capital asset pricing model calculates the tradeoff between risk and return. It's important to know how to use the CAPM because although the impact of current biomedical breakthroughs is exceptional, the risks that come with them are quite significant.

Additionally, the cost of capital in both the biotech and pharma sectors will also be investigated. The two sectors are different in understandable ways. It's therefore essential to disclose some of the reasoning behind the highs and lows of the fluctuating market valuations seen in both industries.

Maximizing Shareholder Wealth

To maximize shareholder wealth is to work in the best interest of the company's investors. Different entities have different views about what it means to maximize their welfare. For some, increased material wealth may only be one dimension and also may not be the most important one. Ordinarily, company objectives and regulations, regarding the company's best interest, should be written out clearly. This way, the company's management team, for example, the corporate manager, the CFO or the CEO, can know explicitly when they are in compliance or in violation of them.

However, legally speaking, the purpose of private sector ventures is to maximize shareholder wealth. Meaning that the management team is legally charged with the responsibility of maximizing shareholder wealth and can therefore be sued for violating the trust of the shareholders if they do not.

The irony is that when dealing with making investments in life-saving drugs, ethical considerations relating to how important maximizing shareholder wealth should be, come to place. Ethical and unethical aspects and applications of finance in the life sciences and healthcare will be looked at in a later chapter of our book. Before that, let's look at the following section which focuses on NPV, that is, maximizing shareholder wealth, as applied to biotech.

NPV Rule and Biotech

The NPV rule states that to maximize shareholder wealth is to look at a sequence of cash flows, analyze their net present value, convert all of those cash flows into a common time period, add up all of the various different conversions, and then calculate the net.

The NPV rule further states that if this net is positive, accept to take the project. If the net is negative, it is recommended to not accept the project. If there are a variety of projects with no limit on how many to accept, it is recommended to only take the ones that are positive. If the variety of projects are mutually exclusive, it is recommended to take the largest positive project.

That's the summary of how to do capital budgeting. Unfortunately, the two issues that complicate this are that what's going on is not always clear and that to figure out if an idea has actual clinical value, a lot of money would have to be spent.

The cash flow profile of a standard biotech startup begins with an idea. Aside from time and effort, equipment is involved as well. For biotech startups, the equipment costs can be remarkably high. A typical biotech company therefore experiences substantial negative cash flow during the first few years.

In that time period, the reward to risk ratio is very low because it's difficult to assess how helpful the developments might eventually be for patients. So, it's quite difficult to receive funding in that stage due to the fact that it's difficult to evaluate the risk.

In order to figure out whether or not a biotech startup should be embarked on, the decision makers must effectively calculate whether all of their cash flows added together will be bigger or smaller than all of the positive cash flows to be received, assuming success. Venture capitalists focus on a very specific portion of the cash growth curve and attempt to reason out how to estimate what the upside is to learn whether or not the option they have to pursue is good or bad.

A recent controversial research report came to the conclusion that a cure is not good business. That it is much more profitable to develop a drug that has to be taken for the life of the patient than to develop a one-time treatment that will cure the patient once and for all.

Over the course of this book, we will show how to develop the tools needed to perform such a calculation. The calculation first requires some terminology and some basic infrastructure that will be delved into later in the book.

Accounting and Cash Flows

Suppose a drug company has developed a vaccine and needs to choose between two strategies for full scale development, assuming r is 5%.

The first strategy is to bring the vaccine into the market within a year. For this, $1 billion is to be invested now and will yield $500 million, $400 million, and $300 million in profits in years one, two, and three.

The second strategy is to bring the vaccine into the market within two years. This will cost $200 million in investments now and $200 million next year for equal returns of $300 million in years two and three.

To calculate which one of the two strategies, if any, should be accepted and taken, the net present value is to be calculated using the appropriate cost of capital. Drawing a timeline to know exactly what the sequence of cash flows is puts the strategies in a clearer perspective.

$$NPV(CF_0, CF_1, \cdots, CF_T) = CF_0 + \frac{1}{(1+r)^1} CF_1 + \cdots + \frac{1}{(1+r)^T} CF_T (3.1)$$

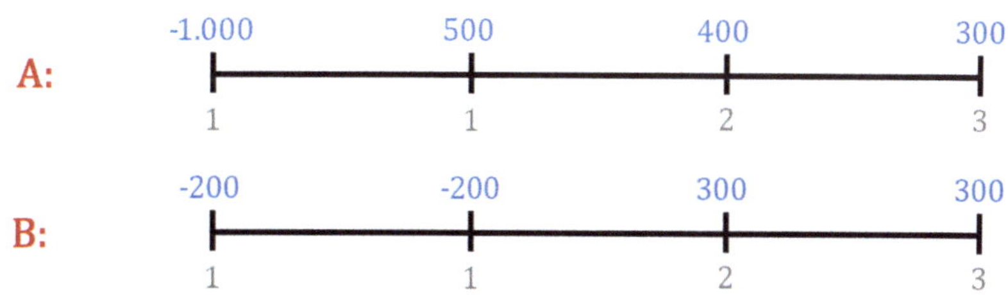

For Strategy A, the net present value is $98 million while for Strategy B, the net present value is $141 million. Since the calculated NPV of the two strategies are positive, it's okay to take and accept both of them. The choice of one out of the two would be to take and accept strategy B With the higher positive net present value.

Table 1 - NPV calculation for Strategy A

Time	0	1	2	3
Cash Flow	-1,000.0	500.0	400.0	300.0
Present Value	-1,000.0	475.2	362.8	259.2
NPV (A)	98.2			

Table 2 - NPV calculation for Strategy B

Time	0	1	2	3
Cash Flow	200.0	-200.0	300.0	300.0
Present Value	200.0	-190.5	272.1	259.2
NPV (B)	140.8			

The main challenge, however, in dealing with a problem like this is not how to calculate the net present value but to analyze the cash flows, which rely on projections. This is where accounting comes in.

In large corporations, the data collated to draw up cash flow figures are in the form of accounting figures. Accounting has a few ways to account for various kinds of cash flows, for example, earnings. The problem is that accounting data are not always a match with market-related variables and need to be translated into net cash flow information.

BALANCE SHEET

Assets	Liabilities
Cash	Equity
Capital	Debt
Intangibles	
Value	Value

The performance of a company can be visualized as the value of what exists within the company at a point in time, as well as how fast that value is growing or shrinking. What exists within a company can be seen in their corporate balance sheets which measure the level of assets or liabilities at a point in time.

INCOME STATEMENT

Sources of Funds = Uses of Funds

On the other hand, how fast the value of what exists in a company is growing or shrinking can be seen in the income statements of the company. An income statement is like a budget constraint for the company that shows the uses of funds in any given year. In the statement, the rate of change of funds has to add up to the sources of funds. It is therefore imperative that the amount of money going equals the amount of money going in.

The basics of accounting prides itself in the fact that the figures always add up because one cannot spend more than they have. Whatever one has is then divide up for use into the various different expenditures.

$$\begin{aligned}
\text{Net Cash Flows} &= \text{Operating Revenues} - \text{Operating Costs w/o Depreciation} - \\
&\quad\ \text{Capital Expenditures} - \text{Income Taxes} \\
&= \text{Operating Profits} - \text{Capital Expenditures} - \text{Income Taxes} \\
\text{Income Taxes} &= \tau \times (\text{Operating Profits}) - \tau \cdot \text{Depreciation} \\
\text{Net Cash Flows} &= (1-\tau) \times (\text{Operating Profits}) - (\text{Capital Expenditures}) + \tau \times \\
&\quad\ (\text{Depreciation})
\end{aligned} \tag{3.2}$$

The conclusion here is that when deciding on whether or not to take on a drug development project it is vital to understand what the project's net cash flows are – which are derived from accounting calculations as seen in the figure above.

Cash Flows vs. Accounting Earnings

Suppose a piece of equipment purchased for $1,000,000 with a life of 10 years generates annual revenues of $300,000 and operating expenses of $100,000. It can be assumed that the equipment will be depreciated over 10 years using straight-line depreciation and that the corporate tax rate is 40%.

Depreciation is an accounting concept that measures how a particular durable purchase gets used up. It reduces the value of an asset over time, because of wear and tear. So, in the accounting books, for example, picture that in operating this $1 million piece of equipment little by little over time, 1/10 of it is used up every year.

Depreciation expense is reported on the income statement as a business expense. The $1 million spent now is an expense and accounting uses the depreciation schedule to allocate the expense across multiple years.

The current earnings measured in accounting is different from the cash that flows in and out of the firm annually. To illustrate this, take a look at the table below.

Table 3 - Earnings and Cash Flows

Date	Accounting earnings before taxes	Accounting earnings after taxes	Net cash flows after taxes
0	0	0	−$1,000,000
1	$300,000 − $100,000 − $100,000 = $100,000	0.6 × $100,000 = $60,000	0.6 × ($300,000 $100,000) − $40,000 = $160,000
2	$100,000	$60,000	$160,000
3	$100,000	$60,000	$160,000
4	$100,000	$60,000	$160,000
5	$100,000	$60,000	$160,000
6	$100,000	$60,000	$160,000
7	$100,000	$60,000	$160,000
8	$100,000	$60,000	$160,000
9	$100,000	$60,000	$160,000
10	$100,000	$60,000	$160,000

The column on the left shows the accounting earnings before interest and taxes. From the first year, the accounting earnings are calculated as revenues ($300,000) minus operating expenses ($100,000) minus the value of the machine used up at the end of the year, 1/10 of $1 million ($100,000). This comes to $100,000, which in the accounting books is the earnings in year 1.

From the perspective of the net cash flows after taxes, as seen in the third column, year 0 writes the $1 million spent as a negative. Starting in year 1, the net cash flows formula can be applied.

Net Cash Flows = $(1-\tau) \times$ (Operating Profits) − (Capital Expenditures) + $\tau \times$ (Depreciation)

Here τ is 0.4

∴ (1 - 0.4) ($300,000 - $100,000) + (0.4 $100,000)
= 0.6 ($300,000 - $100,000) + $40,000 = $160,000
Net Cash Flows is therefore $160,000.

The $40,000 is a depreciation tax yield that is money at the bank due the straight-line depreciation. In this way, accounting has an impact on cash flows. So, accounting is important and cannot be ignored. However, for the purpose of capital budgeting and financing, cash flow figures are more important but do not have the same effect as accounting variables.

For the same equipment, the accounting earnings are $100,000 whereas in cash flow calculations, the depreciation tax yields bring the sum up to $160,000. Therefore, when the net cash flows are calculated and the present value is taken, a different figure is derived than when the accounting earnings are calculated.

The reason cash flow figures are more important than accounting earnings for capital budgeting and financing decision making is because to figure out what something is worth is to focus on what it is about that thing that the stakeholders involved care about the most. As a result, to maximize shareholder wealth is to focus on what investors care about the most. What investors care about the most is cash coming in and going out – cash flows – and the objective is to maximize shareholder wealth.

Practical Considerations

Panacea Medic is a large pharma company contemplating buying off BioCane, Inc., which recently received approval to manufacture and sell its new biologic drug approved for Multiple sclerosis (MS), Alvivorin. Additionally, BioCane has a large variety of other promising treatments in the clinical development stage.

BioCane is registered as a private company and therefore has no market price data on the stock exchange to assess their value. They do, however, have outstanding shares actively held by both external and internal stockholders worth $2.5 million. BioCane also has debt with a market value of $40 million.

To estimate what an appropriate purchase price from Panacea Medic would be, BioCane now wants to make an estimation of what their equity is worth, that is, the value of their shares. To do this, let's assume that BioCane's corporate tax rate is 35% and its cost of capital that will be used to discount future cash flows is 15% annually.

BioCane's management team has made the following projections for its financials (in millions of dollars), for Years 1 through 7, and predicts that its long-run growth rate after Year 7 will be 3% per year in perpetuity.

Table 4 - BioCane Prediction

	Reported				Projected			
Year	**0**	**1**	**2**	**3**	**4**	**5**	**6**	**7**
Sales	100.0	108.0	116.6	126.0	136.0	146.9	158.7	171.4
Cost of goods sold	70.0	75.6	81.6	88.2	95.2	102.9	111.1	120.0
EBITDA	30.0	32.4	35.0	37.8	40.8	44.0	47.6	51.4
Capital expenditure	12.0	15.1	16.3	17.5	15.5	16.2	16.8	16.3
Depreciation	3.2	10.0	10.8	11.2	12.0	12.4	12.9	13.2
Working capital (% of sales)	16.0	17.3	18.7	20.2	21.8	23.5	25.4	27.4
ΔWorking capital		1.3	1.4	1.5	1.6	1.7	1.9	2.0

How can BioCane's share price be calculated based on these forecasts and estimates?

Before solving this example, note that capital expenditures and working capital need to be taken into account to calculate the company's share price. They are part of the initial investment.

Capital expenditures are the pieces of equipment bought and, in some cases, sold. They affect net cash flows and do not have a massive impact on earnings unless under the influence of accelerated depreciation.

Meanwhile, working capital is the liquid cash needed to run the operations of a firm or a company. For example, when creating an inventory of drugs or devices, the many suppliers of the inputs necessary to develop the products need to be paid.

The inputs can then be stored in inventory shelves and are paid for using working capital. If the operations of the firm are shut down, the working capital cash will be retrieved. This is why working capital is taken into consideration when evaluating a particular project.

In BioCane's case, they are a company operating with both debt and equity. Their capital structure needs to be worked from backwards to calculate the worth of their equity. Using accounting to determine net cash flows will be relevant here whereas accounting earnings will not.

The company continues to operate and expand beyond Year 7. Consequently, the different pieces of the company's life must first be consolidated, then the annuity discount formula as well as the perpetuity formula will be used to calculate what the share value is i.e., a combination of what happens between year one to seven and then from Years 8 into infinity.

Table 5 - BioCane Net Cash Flows

	Reported				Projected			
Year	**0**	**1**	**2**	**3**	**4**	**5**	**6**	**7**
EBITDA	30.0	32.4	35.0	37.8	40.8	44.0	47.6	51.4
Capital expenditure	12.0	15.1	16.3	17.5	15.5	16.2	16.8	16.3
Depreciation	3.2	10.0	10.8	11.2	12.0	12.4	12.9	13.2
Working capital (% of sales)	16.0	17.3	18.7	20.2	21.8	23.5	25.4	27.4
ΔWorking capital		1.3	1.4	1.5	1.6	1.7	1.9	2.0
Taxes		7.8	8.5	9.3	10.1	11.1	12.1	13.4
Cash flows		**8.2**	**8.8**	**9.5**	**13.6**	**15.0**	**16.8**	**19.7**

The calculations begin by getting the figures for the projected corporate income taxes. The formula is:

$$\text{Income Taxes} = \tau \times (\text{Operating Profits} - \text{Depreciation} - \text{Amortization}) \quad 3.3$$

The income taxes in Year 1 are therefore:

0.35 ($32.4M $10.0M) = $7.8 million

This formula is then used to calculate the taxes due in Years 2 to 7.

These income tax figures are employed to calculate cash flows. The formula is:

Net Cash Flows = Operating Profits - Capital Expenditures - Income Taxes - Δ Working Capital

The net cash flows in Year 1 are therefore:

$32.4m $15.1m $7.8m $1.3m = $8.2 million

The net cash flows for the next years up to Year 7, and their present values discounted at the 15% per year cost of capital, are amount to the following:

(in Millions)	Year 1	2	3	4	5	6	7
Cash Flow	$8.2	$8.8	$9.5	$13.6	$15.0	$16.8	$19.7
NPV of Cash Flow	$7.1	$6.7	$6.2	$7.8	$7.5	$7.2	$7.4

After Year 7, cash flows are forecasted to grow at a long-term growth rate of 3% per year. Therefore, Year 8's projected cash flow is:

$19.7m 1.03 = $20.3 million

In Year 7, the NPV of the cash flows for Year 8 and beyond is:

NPV at Year 7 =

In terms of a growing perpetuity this can be valued as:

NPV at Year 7 =

Discounting this figure back to year 0 evaluates the NPV of cash flows for years 8 and beyond to be:
NPV at Year 0 =

At this stage, all the elements required to place a value on the business have been calculated. To pinpoint BioCane's equity value is to subtract the value of its debt from the company's NPV. Lastly, to pinpoint the share price is to divide the equity value by the number of shares outstanding.

Value of assets = NPV(future cash flows) = $113.5 million

Value of equity = Assets Debt = 113.5 40 = $73.5 million

Number of shares outstanding = 2.5 million

Price per share = 73.5/2.5 = $29.40

Therefore, BioCane, Inc. should expect a minimum of $29.40 per share from Panacea Medic.

Point 1

The main summary about capital budgeting the standard way is to use cash flows, not accounting earnings, because accounting earnings serve a different purpose than NPV calculations. The cash flows should be after-tax cash flows attributable to the project incrementally.

Some costs need to be taken into consideration, that is, the costs specific to the project addressed. This may prove challenging when dealing with a large enterprise sharing equipment across projects and products. The costs of these shared equipment will have to be distributed proportionately to determine what they are for the specified project. This avoids overestimating the project's value due to a presumption that there are freely available inputs.

Another example can be seen below. In this example, MedCane Tech, Inc. a catheter manufacturer, makes use of various kinds of accounting relations. The ultimate decision should be calculated.

MedCane Tech, Inc. is looking to introduce a new coronary balloon catheter called the Cane IV, which was developed at an R&D cost of $1 million over the past 3 years. A new machine to be used to manufacture the Cane IV would cost $2 million. This new machine will be in operation for 15 years, with a commercial salvage value of $50,000. The machine can also be depreciated linearly to $0 over 10 years with an annual operating cost of $40,000.

The distal shaft of the Cane IV needs to be welded onto the proximal shaft. This welding will be done using excess capacity of an existing plastic welding machine currently operated at a cost of $30,000 (regardless of how heavily it will be operated).

Revenue received will be $400,000 per year, but cannibalization would cause existing sales of MedCane Tech's other range of catheters to decrease by $20,000 per year. Cannibalization is the marketing strategy of a company making a reduction in sales volume, sales revenue, or market share of one product when a new product is introduced by the same company.

The working capital needed over the life of the project is \$250,000. Tax rate is assumed to be 34% while the discount rate of 10%.

Before calculating the NPV, let's take a look at which of these variables should be included in the calculations.

Sunk costs are said to be money already spent that cannot be retrieved. One thing to keep in mind with this example is there are certain sunk costs that should not be included in the valuation calculations. It is crucial to view the project independently. In the case of MedCane Tech, that R&D expense of \$1 million is a sunk cost.

MedCane Tech's initial investment also includes their capital expenditure and working capital. The depreciation is calculated as dividing \$2 million by 10. This puts depreciation at \$0.2 million for the first 10 years. The machine's salvage value is fully taxable since the book value at the end of year 10 is \$0 when the machine cost will have been fully depreciated.

Additionally, the project should not be charged for time used on the plastic welding machine but should be charged for the cannibalization of the other catheter sales figures.

Years	Cash flow (\$1000's)		
0	−(\$2000 + \$250)	=	−\$2,250.0
1-10	(\$400 − \$40 − \$20) × (1 − 0.34) + \$200 × 0.34	=	\$292.4
11-14	(\$400 − \$40 − \$20) × (1 − 0.34)	=	\$224.4
15	\$224.4 + \$50 × (1 − 0.34) + \$250	=	\$507.4
NPV:	−\$57.62		

Point 2

The other summary about capital budgeting is about project interactions. Projects can interact in two ways, deciding among a set of projects and deciding the optimal timing of projects.

If projects are standalone, the NPV rule can be applied to them. However, if projects are dependent, this dependence has to be taken into account as specifically as possible. For example, if they are mutually exclusive and accepting one rules out the others, their NPV's can be compared to make a decision.

With regards to the optimal timing of a project, the options are to reject the project completely, accept the project now, accept the project later or to wait until the NPV increases. Suppose there's a particular product with a value that increases over time because the demand for it increases the longer the market awaits its launch. If the project is started early, competitors might catch up by replicating the product. The opportunity cost of capital, the loss of other alternatives when an alternative is chosen, is 10%.

Now, the question of when the optimal time to launch the product will be, comes to mind. There's the chance that the project will have a higher valuation by allowing the market to grow while waiting – but at the same time, there's no turning back after launching the product.

By calculating the future present value of the project if it were launched next year or the year after that, and so on, and bringing it all back to present value, the following table can be drawn:

Year to Start	FPV	% Change in FPV	NPV
1	100	-	91
2	120	20	99
3	138	15	104
4	149	8	102

According to the table, Year 3 symbolizes the point where the growth rate of the present value equals the growth rate of the profitability of this market. This is under the assumption that the growth rate stays below 10% afterwards.

What this means is that before year 4, the return for waiting is larger than the opportunity cost of capital. Since the growth rate of FPV remains below 10% after year 4, the best suggestion is to wait and introduce the product at the end of year 3.

Next, let's look at alternatives to the NPV.

Alternatives to NPV

In practice, other investment rules aside from NPV are also used in capital budgeting. These include the Adjusted Present Value (APV), Profitability Index (PI), Internal Rate of Return (IRR), Price-to-Earnings (P/E) Multiples, Accounting Rate of Return (ARR), real option analysis, sensitivity analysis, hurdle rate, value-at-risk simulation analysis, payback period, discounted payback period, and other techniques.

Companies apply these rules to the valuation of projects because they may have worked for the decision making encountered by them in the past. Every so often, these rules derive at similar decisions as the NPV but in general, they do not. At the end of the day, the NPV rule prevails over these alternatives. It is important to stay vigilant of the disadvantages of relying on NPV alternatives and to therefore use NPV whenever possible.

The IRR is beneficial when assessing biotech venture capital financing because although the NPV yields internally consistent provisions for what a project is worth, risk is involved. NPV offers the assurance of knowing what cash flows will be in a world without risk whereas investors have to apply due diligence to avoid career damaging decision making. As a result, project valuations need to take into consideration a range of additional factors and features.

Internal Rate of Return

Let's review the Internal Rate of Return (IRR). Here, the main idea is to calculate, for a given set of cash flows, the return that an investment is going to yield.

$$I_0 = \frac{CF_1}{(1+\text{IRR})} + \frac{CF_2}{(1+\text{IRR})^2} + \cdots + \frac{CF_T}{(1+\text{IRR})^T} \quad (3.4)$$

Picture this: the day the initial investment is made is day 0. Then there are cash flows starting in the next time period. This is given as CF1 through CFT. The IRR calculation sees that the initial investment adds up to all of the cash flows. In other words, it calculates the single number that arrives at a zero NPV figure.

The IRR can also be viewed in the same way as the yield to maturity of a standard coupon bond – as the yield to maturity of a project. Similar to the average return that such a sequence of cash flows offers its investor. So, instead of coupons and the principle as in a bond, the IRR deals with the cash flows of the project.

When concluding on project valuation decisions based on the IRR, two criterions can be followed. For independent projects the suggestion is to accept a project if its IRR is greater than the threshold rate (the minimum rate of return that an investor can expect to receive after investing in a project). On the other hand, for mutually exclusive projects, accept the project with the highest IRR among the projects with an IRR greater than the threshold rate.

There are four instances where the IRR rule leads to the same decisions as the NPV. The first instance is if there is only one cash outflow, which occurs at time 0, the day of the initial investment. The second instance is when there is only one project under consideration. The third instance is when the opportunity cost of capital is the same for all time periods. The last instance is when the threshold rate is set to be equal to the opportunity cost of capital.

The IRR is the primary measure used by venture capitalists to calculate the rates of return on their startup investments. This is because the sum total of how much money they'll put in and get out of a project is oftentimes the only piece of data they have to work with for valuation purposes. For their calculations, investments put in later on are not calculated as part of their initial investment – but calculated as different investments analyzed on their own.

For startups, it's quite challenging to calculate rates of return any other way because startups neither trade on the stock exchange nor generate cash flows initially. As a result, it will be difficult to pinpoint exactly what a single biotech project will be worth. At any point in time, the project may be faced with illiquid markets or distressed sales so standard net cash flow NPV calculations cannot be applied.

There are a special set of circumstances when the IRR will not lead to the same decisions as the NPV. It is therefore important to be careful in the cases where no IRR can be computed and in cases where there will be multiple IRRs.

There are a special set of circumstances when the IRR will not lead to the same decisions as the NPV. It is therefore important to be careful in the cases where no IRR can be computed and in cases where there will be multiple IRRs.

If a project involves cash flows that come in initially and go out later on, similar to a loan received and paid back out with interest, the cash flow is said to be complicated. A practical biotech example of this is when funding is poured into a phase one clinical trial. Then after the phase reaches a milestone, more funding is poured in for phase two of the trial. An IRR cannot be found for cases like this so the IRR method cannot be used for valuation purposes.

Similarly, arbitrarily deciding on multiple IRRs will lead to the wrong answer.

Funding Case Study: Genentech & Herceptin

This section will discuss Genentech, an industry pioneer in biotechnology as well as Herceptin, a monoclonal antibody specifically treating HER2 receptor positive metastatic breast cancer.

Genentech

Prior to the 1970s and the revolution of biotechnology, drug development processes were more focused on chemistry, and not biology. The main approach was to target a disease, come about a combination of chemical compounds and then test them on patients. As long as the drug treated a fraction of patients, it was approved with the appropriate warning labels.

A different approach to developing drugs was however introduced in the 1970s, '80s, and '90s i.e., to use biology, not chemistry, to create drugs. The new approach focused on two things.

Chemical compounds are synthetically manufactured in a lab whereas biological agents are derived from living things. So, firstly, the new approach was geared towards developing biological agents instead of continuing to inject humans with chemical compounds as a form of treatment.

Secondly, the new approach did not apply the trial-and-error testing of chemical compounds to see which drugs worked and which ones did not. It instead targeted the specific biological mechanism of a disease and worked to challenge that mechanism. Genentech was a pioneering member of this new approach and a part of that revolution

The company Genentech was founded by venture capitalist Robert A. Swanson and biochemist Herbert Boyer in 1976. Swanson served as CEO of the company from 1976 to 1990, and as chairman from 1990 to 1996 while Boyer is regarded as one of the founding fathers of the recombinant DNA technology field.

In the mid-70's recombinant DNA was not an existing widespread tool and the science around it was in its infancy stage. The first publications describing the successful production and intracellular replication of recombinant DNA appeared in 1972 and 1973, from Stanford University and University of California, San Francisco (UCSF), the latter being Boyer's alma mater. So, recombinant DNA technology was the company's unique focus at the time that they established the company in 1976.

The idea behind recombinant DNA (rDNA) is that DNA molecules can be mixed and matched with genetic material from multiple sources to create compounds that do not exist in nature.

For example, creating human insulin by inserting the appropriate instructions into the DNA of bacteria, like E. coli or yeast.

The first approved product developed by Genentech was their synthetic human insulin, Humulin, in partnership with insulin manufacturer, Eli Lilly and Company. The bacteria E. coli was used with the suitable segment of DNA inserted into it to create human insulin for diabetic patients. Upon FDA approval, Humulin was the first approved genetically engineered human therapeutic. That was also historical for the biotech Industry because at the time, drugs were not developed and approved for biotech companies.

Genentech went public in 1980 and their IPO raised $35 million, equivalent to approximately $150 million in the third decade of the 21st century, using BRDPI. In the first hour of trading, Genentech's IPO price rose from $35 a share to $88 a share. This was the beginning of the biotech industry.

To earn more stable cash flows, the company made a deal with F. Hoffmann-La Roche AG (Roche), a Swiss multinational Holding healthcare company. The deal saw that in 1990, Roche acquired majority ownership of Genentech and the acquisition was completed by 2009.

As at the time of writing this book, Genentech does not trade on the New York Stock Exchange or NASDAQ anymore because it trades under Roche. In the initial partnership, Roche did not exercise their control over Genentech so up until 2009, Genentech traded separately and independently of Roche on the stock market.

Following the merger, Genentech sits as the headquarters for all Roche pharmaceutical operations in the United States. As a result, products marketed by Roche in the US are now marketed by Genentech in the United States. The combined portfolios of both companies allow for the concentration on the use of human genetic information to develop treatments that serve for patients with serious or life-threatening health conditions.

The timeline to the left shows the treatments approved in the United States by the combined company. To the left are treatments under Roche while to the right are treatments under Genentech. It is evident that the merger became a powerful force behind new drug development through shared intellectual property and technologies.

Genetech's founders, Swanson and Boyer, originally set out to develop drugs differently than it was being done before the 1970's. However, that was a challenge – especially to push their drugs all the way through to approval and marketing. Whereas Roche's financial backing contributed to relieving the challenge.

The deal came with a bit of controversy after it was announced in 1990. As part of the acquisition, Roche injected approximately $400 million into Genentech and gave way for access to more pharmaceutical development resources. Unfortunately, this was viewed by some as a lack of integrity on Genentech's side. The term used by The New York Times was that they were sell-outs.

On the other hand, some scientists appreciated the deal for providing more funding for research. They viewed it as a shield from the capitalist pressures that had been forcing the company to restrict some of their researchers' freedom.

The degree of independence that Genentech would have also been questioned. In the end, however, the story worked out well. Genentech continues to have a separate identity from Roche and remains an innovative company.

Nonetheless, and beyond individual opinions, the financial pressures mattered because left to their own devices, the volatility of how Genentech performed on the stock market over the years would have made it difficult for them to raise capital in case their research needed it abruptly. Since scientific breakthroughs are not correlated with market fluctuations, it was important for Genentech to maintain stable cash flows and funding. Ultimately, they had to make a decision.

Herceptin

Herceptin is the trade name of a treatment, developed by Genentech, for breast, stomach and esophagus cancers that are HER2 positive i.e., positive for the human epidermal growth factor receptor 2 protein that causes cancer cells to grow. The treatment's generic name is Trastuzumab and is an artificially created antibody that fights off cancer cells, specifically the HER2 protein, similar to the way naturally developed antibodies fight off germs.

Herceptin can be used for treatment in three ways. The first way is called the adjuvant treatment. This is done after the cancer has been surgically removed. The second way is called the neoadjuvant treatment which is done before surgery in order to shrink the cancer. This makes it easier to remove the cancer by surgery. The third way is called advanced or metastatic cancer treatment. This is for cancer that has already spread and cannot be removed by surgery.

At certain periods during the development of the drug, Genentech cut off funding and their support. Some fundraising from the philanthropic sector had to be done in order to raise money to continue with the R&D process. It took a number of years of basic science before even a handful of those scientific discoveries were applied in the clinic and on patients.

The Herceptin development timeline starts back in 1975 when the potential of using antibodies in vitro to fight disease was discovered. The discovery was around the idea that there was a genetic component to how cancer tumor cells grow, that those genes could be looked at, and ways to interfere with them could be developed. This was the design concept behind Herceptin and at the time, it was a revolutionary approach to dealing with cancer.

The drug's basic scientific research stage extended from 1975 to 1984. During this time, a Nobel Prize in medicine was won for 'theories concerning the specificity in development and control of the immune system and the discovery of the principle for production of monoclonal antibodies. The preclinical stage as well as Phase I and II of the clinical trials were initiated and completed between 1985 and 1994.

By 1998, Herceptin became the first therapeutic antibody targeted to the HER2 cancer molecular marker to receive FDA approval. Genentech and Roche also signed a licensing agreement giving Roche exclusive marketing rights outside of the United States for Herceptin.

Part of the reason the development process was completed as fast as it was, was because their approach particularly showed substantial efficacy. There are patient populations that have been administered Herceptin along with other chemotherapy, radiation therapy, surgery and/or combination therapies that have been cancer free for up to 20 years.

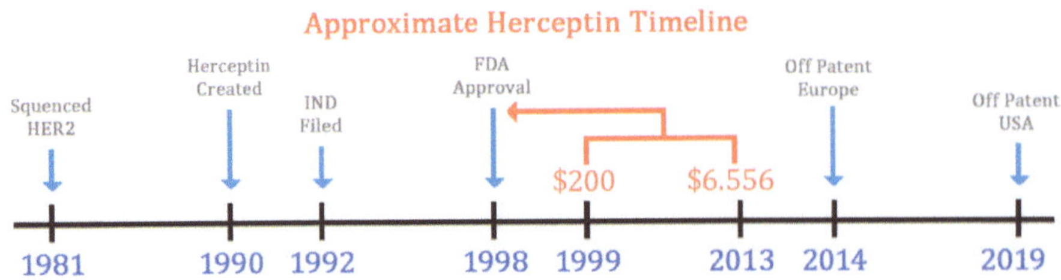

Figure 7: Herceptin Timeline

To analyze whether the product, Herceptin, was a good investment, it's important to ask who the investors were as well as what their objectives, constraints, and timelines were. Additionally, although there are a lot of different stakeholders in biomedicine, this section will only focus on investments from the shareholder perspective. So, the most important question to ask is about what the present value of all of the earnings is.

To calculate the NPV of this product the development timeline will need to be broken down. It can start at the time of sequencing the HER2 gene i.e., 1981. Then Herceptin was created in 1990 and the clinical trials began in 1992. They received FDA approvals in 1998 so the drug began to sell around 1998 to 1999. The first year of sales brought in $200 million and the figures kept rising. In 2014, the European patent expired, and the US patent expired in 2019. These must be taken into account because when a drug's patent protection expires, some market share will be lost to generic drugs.

From 2014 to 2019, the European patent had expired, but the US patent was still active. Revenues from then can be totaled and computed. The final time period is after 2019, when both the European and US patents had expired and it can be assumed that because of that, sales figures decreased by 20%. This is how the NPV calculations can be broken down. In reality, other factors such as legal risk, scientific risk, business risk and financing risk will need to be considered as well.

Using the actual realized sales figures for Herceptin to calculate the net present value in US dollars as of 1998 comes up to $19 billion at a 10% cost of capital. Using their projected revenue figures, as at today, Herceptin is worth approximately $54.6 billion in NPV, again at a 10% cost of capital, in addition to how much it was worth in 1998. The calculations conclude that not only is Herceptin a good investment but also, the lives of hundreds of thousands of women have been positively affected by this drug.

Combination therapies have revitalized Herceptin and massively contributed positively to its enduring success. In 1998 at the time of approvals, this could not have been anticipated to forecast cash flow figures. One such contributing factor to its success is the approval of a new drug by

Genentech called Perjeta. Similarly to Herceptin, Perjeta is a monoclonal antibody therapy but targets a slightly different mechanism. When used in conjunction with Herceptin, the combination proves to be more effective than either drug on its own.

Furthermore, when a chemotherapy like Paclitaxel, which hinders cell division, is added, the results are even more promising. Although paclitaxel is a generic drug, the combination of all three treatments appears to be the most effective. These factors greatly influence the sales figures, making it challenging to accurately forecast them. Therefore, it is important to approach sales forecasts with caution due to the potential for significant misestimations.

Upon reviewing these figures and considering the expenses associated with research and development, investing in Herceptin appears to be a highly profitable endeavor at first glance. While this may not hold true for all drugs, it tends to be the case in the field of oncology for various reasons. The oncology market allows for significant margin of error due to its profitability, necessitating careful consideration when evaluating different opportunities.

There are valid arguments both in favor of and against these types of incentives. Nevertheless, the current reality is that these incentives do exist. The question then becomes whether the initial investment justifies the potential returns. Based on a wide range of capital costs, it certainly seems to do so. The industry standard of using a 10% cost of capital is employed here, although it may not necessarily be the most accurate figure. If we were to consider a 40% cost of capital, which is an exceptionally high and unrealistic number (typically, biotech companies use 20% or 30%), the projected profits from 2017 onwards would still amount to a remarkable $17 billion.

Determining the appropriate cost of capital is a topic that will be discussed in the upcoming chapters. However, when considering the range of cost of capital for the biotech and pharmaceutical industries, a blend between 10% and 20% seems reasonable. In this particular case, given that Genentech is an intermediary company, a cost of capital of around 15% may be appropriate. However, it is important to note that Genentech is now part of Roche, a major pharmaceutical company. Therefore, a more suitable cost of capital may be around 11% or 12%, as it takes into account bird pie inflation rather than standard inflation.

Despite how profitable Herceptin appears, in reality it encountered significant obstacles that nearly prevented its realization. Genentech cut funding due to their lack of confidence in the drug's potential, but UCLA researcher Dennis Schliemann had to advocate for it.

Dr. Larry Norton, a prominent cancer specialist, also played a crucial role in persuading Genentech to continue with the drug by personally traveling to California at his own expense. Without the efforts of individuals like Dr. Norton, Dennis Schliemann, and philanthropist Ron Perelman, Herceptin would not have been developed fully or approved. Although the drug is remarkable, the journey from the preclinical phase to its approval was far from smooth sailing.

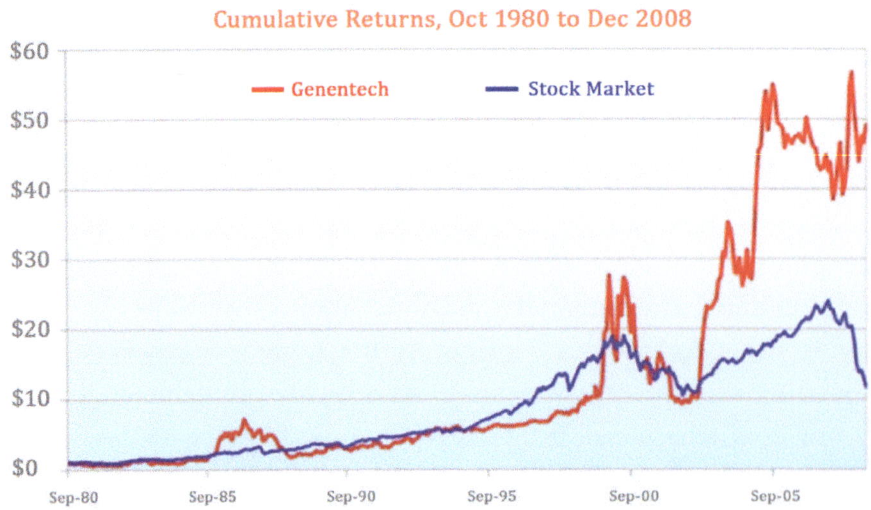

Figure 8: Genentech Cumulative Returns

Even with the support of Roche and Genentech's substantial resources, there were instances where it was considered to discontinue Herceptin due to its high risk. This challenge prompts the contemplation of alternative approaches to streamline the financing process.

As demonstrated by the fluctuations in Genentech's equity growth curve (Figure 8) during their time trading on the stock market, shareholder and investor focus on earnings can greatly impact a company's share price. Consequently, corporate financial executives resort to cost-cutting measures during downward periods. This highlights the challenge often referred to as the "valley of death".

It is imperative that more effective methods of drug development are explored, as there may be numerous other drugs with equal or greater efficacy that lack adequate funding. Manufacturer's peak sales estimates, typically projected five to ten years from launch, have little correlation with actual sales figures. It is unrealistic to forecast anything with certainty over such a lengthy time frame, including the personal preferences of populations. Although forecasting difficulties do not deter investors willing to take risks, it is crucial to identify the right investors for such ventures.

4. STOCKS AND BONDS

Investing in research and development (R&D) clinical trials, as well as acquiring new plant and equipment, often requires significant financial resources. While some companies can use their retained earnings to cover these costs, many need to seek capital from external investors.

Typically, long-term investments are funded by issuing stocks, but companies also have the option to borrow funds for a specific period. This borrowing can be done through bank loans or lines of credit, or more formally through the issuance of bonds, which are also known as fixed income securities.

Bonds offer fixed interest and principal payments throughout their lifespan, making them relatively straightforward to understand in terms of cash flow timing and magnitude. However, the presence of risk and uncertainty adds complications to their evaluation.

On the other hand, stocks can be seen as a continuous stream of cash flows in the form of dividends, stock dividends, or share repurchases. Unlike bonds, these cash flows are uncertain in both timing and amount, as they depend on the company's performance and pay-out policy.

This chapter begins with an overview of the stock and bond markets, followed by a conceptual framework for estimating security prices using present value relationships. The chapter then delves into the increasing significance of debt in the biopharmaceutical industry.

Overview of Stock and Bond Markets

In the wider context of capital markets, although the stock market holds significance, it may not be the most important aspect of the capital market. This concept can be exemplified through a straightforward illustration. The chart on the left side portrays the magnitude of the stock market.

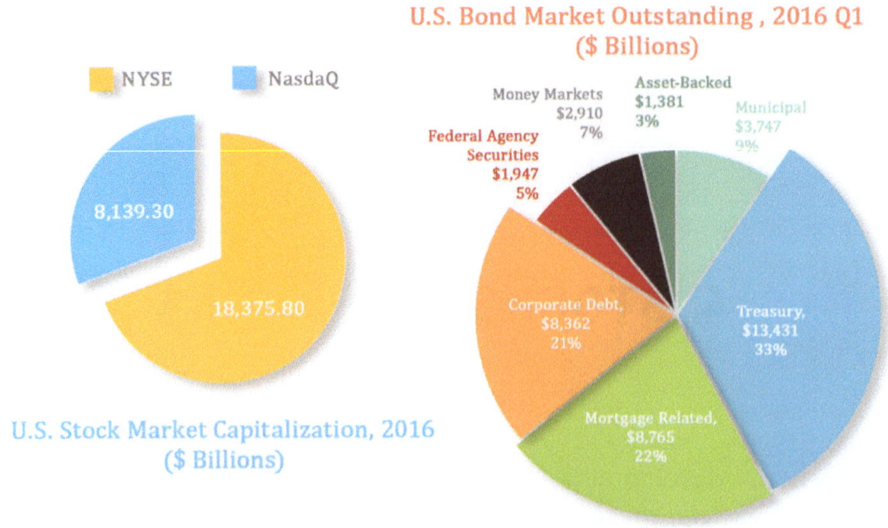

Figure 9: US Stock and Bond Market 2016

In 2016, the NASDAQ possessed a total value of approximately $8 trillion, while the entire stock market amounted to $18 trillion. Although these figures are substantial, they are notably less in comparison to the bond markets, which reached roughly $39.9 trillion during the same period. The bond markets, including solely the corporate debt market, surpass the value of all stocks on NASDAQ.

Moreover, mortgage-related debt totaled $8 trillion, while asset-backed securities amounted to $1.3 trillion. These figures highlight the large scale of these markets. When considering the issuance of equity and debt, the discrepancy between the two becomes even more pronounced. In 2016, initial public offerings (IPOs) raised $17.8 billion, whereas the issuance of corporate bonds raised $1.5 trillion.

This significant disparity emphasizes the contrasting nature of equity and debt markets. While equity markets entail higher risk due to the sale of a stake in a company's future, they offer the advantage of permanent capital. Conversely, debt markets provide a larger pool of funds that can be utilized for various purposes, such as drug development.

Exploring innovative approaches to leverage fixed income markets is crucial, as it has the potential to revolutionize the industry. However, it is important to note that the current industry practices do not align with the approach that will be discussed in subsequent chapters. Therefore, adopting a forward-thinking mindset and demonstrating courage will be necessary to implement these changes.

This book will showcase different chapters that address this topic, encouraging readers to consider utilizing fixed income markets in unprecedented ways.

The Pricing of Stocks

In the context of stocks, the approach is similar to that of other financial instruments. However, stocks differ in that they represent ownership in a company and do not offer fixed cash payments.

Unlike bonds, the cash flows from stocks are uncertain and the duration of payments is indefinite. Stockholders are considered permanent owners of the company's cash flows, which necessitates an understanding of certain factors to determine the value of the stock.

To value a stock, one must assess the cash flows it generates. The value of a stock today is determined by the present value of all future cash flows, which are expressed as dividends received by shareholders. However, predicting these cash flows is challenging as they are subject to randomness. Unlike bonds that have predetermined coupon payments, the cash flows from stocks are uncertain.

Additionally, the valuation of stocks requires the consideration of exchange rates. When incorporating risk into the valuation process, one must account for how to compensate for that risk. This is where the capital asset pricing model (CAPM) comes into play. The CAPM helps determine the appropriate discount rate to adjust for the risk associated with a particular company.

Once the cash flows and risk adjustment figures are known, the pricing of bonds and stocks becomes similar. A simple approach to valuing stocks involves considering a reliable company that pays dividends that grow over time at a rate denoted as . The formula for valuing such stocks is expressed as:

$$PV = \frac{C}{r-g} \quad , \quad r > g \tag{2.8}$$

Where:

C = Expected future dividends

r = Discount rate based on the CAPM

g = Growth rate

Understanding the factors that affect stock prices is crucial. Three variables influence the formula mentioned above: cash flows (dividends), the discount rate, and the growth rate. Estimating future cash flows requires leveraging collective wisdom and expertise. The discount rate is determined by assessing the risk associated with the company. Lastly, the growth rate, g, plays a significant role in stock valuation. Any small change in g can have substantial impacts on the value of the company.

To illustrate the importance of g, consider the example of a company with a 10% cost of capital and a 5% growth rate. If the company pays a $1,000 dividend per period, growing at a rate of 5% would yield a stock value of $20,000. However, if the growth rate is 0 instead of 5%, the company's value is halved.

The difference between a 5% and 0% growth rate is significant, especially when the cost of capital and growth rate are close together. Even small changes in these variables can lead to substantial fluctuations in the company's value.

This information is relevant in the context of biotech companies, as they constantly assess future cash flows and prospects. These companies face binary risks, where a negative outcome in a clinical trial can have disastrous consequences. For instance, if the growth rate changes from 5% to 0%, the value of the company is greatly affected.

Conversely, positive outcomes can lead to significant gains in stock value, as exemplified by Puma Biotechnology. When the results of their phase III clinical trials for a cancer drug called Neratinib were positive, the company's stock price skyrocketed from \$59 to \$233 per share, resulting in a one-day return of \$1 billion for a hedge fund.

On the other hand, companies can experience severe losses due to negative clinical trial outcomes. For example, when the company Solid Biosciences' gene therapy for Duchenne muscular dystrophy failed to produce the desired results in phase II trials, the company's market cap plummeted by 76% in a single day. This example highlights the extreme volatility associated with gene therapies and the high level of risk involved. Managing financial risks and considering scientific risks becomes crucial in such circumstances.

Valuing the assets of companies requires analyzing the various stages of cash flows. This process is complex and necessitates an understanding of the industry and business. Different estimates and discount rates must be employed in order to accurately assess the value of such cash flows.

It is a fundamental principle that the cost of capital cannot be lower than the growth rate and a core concept in equity valuation is to meticulously calculate the growth rates for different stages in a company's life cycle. This involves breaking down the company's future into distinct stages, namely the growth stage, transition stage, and mature stage, rather than relying on a simplistic dividend discount formula that assumes dividends do not change or grow perpetually at the same rate.

The growth stage sees rapidly expanding sales, high profit margins, high growth in earnings per share, multiple new investment opportunities and a low dividend payout ratio. The transition stage sees the growth rate and profit margin reduced by competition, fewer new investment opportunities and a high payout ratio. The maturity stage sees an earnings growth. Furthermore, in this stage, the payout ratio and average return on equity stabilizes for the remaining life of the firm.

Various formulas can be employed for each stage, such as the perpetuity formula for the mature stage, the annuity formula for the transition stage, and the annuity formula with growth for the growth stage. By piecing together these components, one gains a comprehensive understanding of the company's true value. This approach forms the foundation of equity valuation for such investments.

While the concept may appear straightforward, executing it effectively necessitates a profound understanding of the industry. Financial analysts typically possess detailed knowledge of factors that impact variables such as discount rate, growth rate, dividends, the time horizon, the drug pipeline, FDA regulations, and pricing pressures within companies. Analysts can also provide comprehensive insights into these considerations.

That concludes the coverage of equity valuation. Going forward, we will apply various aspects of this concept to examine valuations within the biopharmaceutical industry.

Bonds

The focus now shifts towards bonds and what are considered the most crucial concepts pertaining to bond pricing. The fundamental notion of a bond is likely familiar to all. It represents a fixed income claim, implying that the income can be predetermined and is not subject to randomness. Prior knowledge enables one to ascertain whether the income will be derived from coupon payments or the final repayment of the principal.

Additionally, bonds possess a specific maturity date. For instance, a three-year bond with a face value of $1,000 serves as an illustrative example. It offers an annual coupon payment of 5%, resulting in three payments of $50, disbursed at the end of each year. Furthermore, at the conclusion of the third year, the principal amount is also repaid.

An arrangement like this entails a clear and certain sequence of cash flows, devoid of any uncertainty. The pricing of bonds encompasses various components that warrant emphasis. These include concepts such as the time value of principal and coupons, which is essentially the net present value (NPV) as well as exchange rates.

Furthermore, the incorporation of risk is imperative. Although an in-depth analysis of risk will be postponed until subsequent chapters, a brief discussion regarding the disparities between different types of bonds will be undertaken. Specifically, there are bonds with inherent risks, which may result in the non-repayment of the principal amount, or default. However, at this moment, the focus will be on riskless bonds, in terms of default.

Although these bonds still entail risks associated with inflation and fluctuating interest rates, one can be assured of receiving the promised nominal dollar amount. The exact purchasing power of these dollars may be uncertain due to factors such as inflation or other considerations.

Two categories of bonds exist: pure discount bonds and coupon bonds. Pure discount bonds solely offer a fixed payment upon reaching the maturity date, whereas coupon bonds provide periodic payments throughout the bond's duration. Zero coupon bonds are constructed from treasury securities and are commonly referred to as strips. These bonds essentially strip away the coupons, treating each coupon as an individual discount bond.

Discount Bonds

Discount bonds are also known as zero-coupon bonds or STRIPS (Separate Trading of Registered Interest and Principal Securities). They trade at a "discount" to face value which means that they do not pay out regularly scheduled coupon payments but make a single payment of principal at maturity. Valuations for discount bonds is a straightforward application of NPV.

$$P_0 = \frac{F}{(1+r)^T} \tag{4.1}$$

The spot rate of a bond is the current yield for a given term. It increases as the term increases so bonds with longer maturities typically have higher yields. If rate varies over time, different interest rates from one year to the next are used and the spot rate of interest in Year T is denoted as . Today's T-year spot rate is calculated as an average of one-year future spot rates. That is, the entire sequence of future spot rates today is not observed.

$$P_0 = \frac{F}{(1+R_1)(1+R_2)\cdots(1+R_T)}$$

$$= \frac{F}{\left(1+r_{0,T}\right)^T} \quad , \quad r_{0,T} \equiv \text{Today's } T\text{–Year Spot Rate} \qquad (4.2)$$

The forward rate for a bond is calculated as a comparison of the anticipated future yield of two bonds. The forward rate represents the yield that would be obtained if the funds from the earlier maturing bond were reinvested to align with the maturity period of the later maturing bond.

The formula for the one-year forward rate for a two-year bond is:

$$/\{[(1+\text{Rate}_{n1})^{n1} \div (1+\text{Rate}_{n2})^{n2}] \div [(\text{Rate}_{n1} - \text{Rate}_{n2})]\} - 1/\times 100 \qquad (4.3)$$

Where:

Rate_{n1} = Spot rate for the two-year bond
Rate_{n2} = Spot rate for one-year bond
n1 = Number of years for the first bond
n2 = Number of years for the second bond

In the context of a forward transaction to obtain a loan in the future, the terms of the transaction are established at the present moment, denoted as t = 0. The actual loan is received on a subsequent date, referred to as t_1, while the repayment of the loan takes place on a later date, denoted as t_2. It is important to note that the future spot rates may differ from the current forward rates.

Bond futures rates refer to the expected future interest rates on bonds. These rates are derived from the prices of bond futures contracts traded on exchanges. Bond futures contracts allow investors to speculate on the future direction of interest rates and hedge against interest rate risk. The futures rate represents the market's expectation of where interest rates will be at the time of the bond's delivery.

Suppose several discount bond prices are observed today:

$$\begin{aligned}
P_{0,1} &= \frac{F}{(1+R_1)} \rightarrow r_{0,1} \\
P_{0,2} &= \frac{F}{(1+R_1)(1+R_2)} \rightarrow r_{0,2} \\
P_{0,3} &= \frac{F}{(1+R_1)(1+R_2)(1+R_3)} \rightarrow r_{0,3} \\
&\vdots \\
P_{0,T} &= \frac{F}{(1+R_1)(1+R_2)(1+R_3)\cdots(1+R_T)} \rightarrow r_{0,T}
\end{aligned} \tag{4.4}$$

$$\{P_{0,1}, P_{0,2}, \cdots, P_{0,T}\} \rightarrow \{r_{0,1}, r_{0,2}, \cdots, r_{0,T}\}$$

Term Structure of Interest Rates

The term structure of interest rates refers to the correlation between interest rates or bond yields and various terms or maturities. When depicted graphically, the term structure of interest rates is commonly referred to as a yield curve, and it assumes a vital significance in discerning the present condition of an economy.

In summary, the current interest rate for a loan that is immediately available is referred to as the "today's spot rate". This rate is determined and recorded as $r_{0,1}$, $r_{0,2}$, and so on, as of the present day (date 0). The current interest rate for a loan that will be taken out in the future is known as the "forward rate." This rate is also determined and recorded as f_2, f_3, and so on, for loans with a duration of one period. However, for a loan that will begin in year 3 and last for a duration of 4 years, it is denoted as $f_{0,3,7}$.

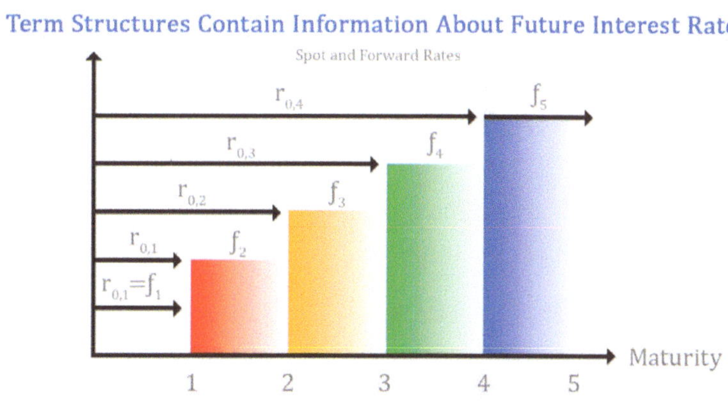

Figure 10: Term structure spot and forward rates

The rate for a future loan, which is not known at the present time, is called the "future spot rate". This rate is unknown as of today (date 0) and is represented as R_1, R_2, and so on, for future one-period loans. However, for a loan that will begin in year 2 and last for a duration of 5 years, it is denoted as $r_{2,7}$.

The following example demonstrates the utilization of the disparities between forward and futures rates, as well as spot rates.

Consider a hypothetical scenario where one assumes the role of a chief financial officer in a prominent pharmaceutical company. In this scenario, the CFO will require a loan of $20 million in one year's time for a three-year phase II or phase III trial. The objective is to determine the cost of obtaining this loan and explore the possibility of securing a favorable interest rate in the present.

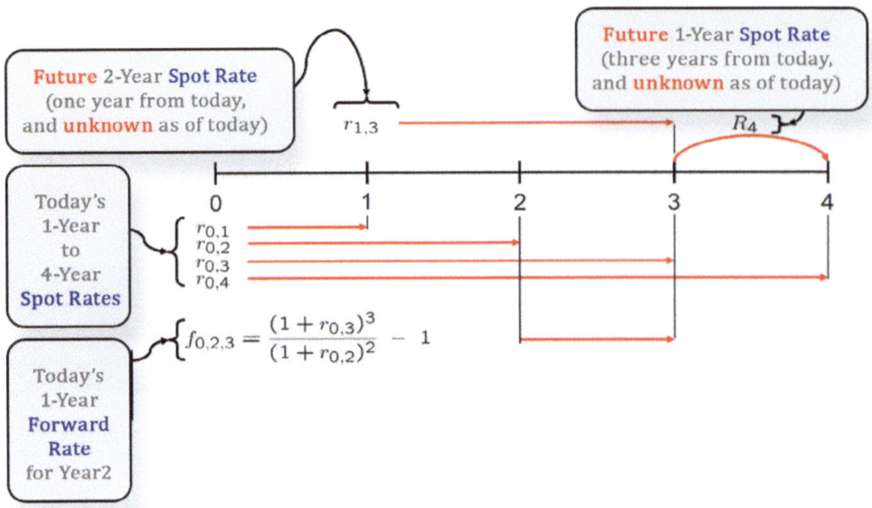

Figure 11: Future Forward Rate formula

To ascertain the potential cost and feasibility of securing the loan, the CFO must consider the risks associated with waiting for a year, as interest rates may fluctuate unfavorably during that time. By locking in the current interest rate, the CFO can mitigate the risk of facing exorbitant borrowing costs in the future. Thus, it becomes imperative to evaluate whether it is advisable to secure the rate now based on the CFO's anticipation of interest rate movements.

Suppose that discount bond prices are as follows:

t	1	2	3	4
P_t	0.9524	0.8900	0.8278	0.7629
$r_{0,t}$	0.05	0.06	0.065	0.07

To facilitate this evaluation, the CFO examines the yield curve or term structure of interest rates, specifically the spot rates available today. The spot rate for borrowing or lending over one year is 5%, while it is 6% for two years, 6.5% for three years, and 7% for four years. This term structure indicates an upward trend in interest rates, as borrowing for longer durations necessitates higher rates. This aligns with the CFO's belief that interest rates will increase.

The CFO's objective is to determine the future three-year spot rate one year from now, which is currently unobservable. Since it is impossible to predict the future with certainty, the CFO and other individuals involved cannot access the exact future three-year spot rate.

The forward rate for a three-year loan one year from now can be determined by examining the difference between the four-year rate and the current one-year rate. This can be calculated by normalizing the ratio of the four-year spot rate to the one-year spot rate. In this case, the implicit forward rate is found to be 7.68%. This implies that a \$20 million three-year loan taken out next year would require a payment of 7.68% interest. This rate can be locked in using a specific term structure of interest rates and access to bonds.

The Year-1 three-year forward rate $f_{1,4}$:

$$(1+f_{0,1,4})^3 = \frac{(1+r_{0,4})^4}{(1+r_{0,1})} = \frac{(1.07)^4}{1.05}$$

$$f_{0,1,4} = \left(\frac{(1.07)^4}{1.05}\right)^{1/3} - 1 = 7.68\%$$

To lock in the rate, a strategy can be employed. Assuming the bonds have a face value of \$1, the individual is advised to purchase 20 million one-year bonds at the present time. These bonds will mature in a year and pay \$1 each. The cost of purchasing 20 million of these bonds is approximately \$19,047,619, considering that each bond is priced at around \$0.95.

(200,000,000)(0,9524) = \$19,047,619

Simultaneously, the individual should engage in short selling four-year bonds. The exact quantity of four-year bonds to be short sold is not specified, but one can assume that 24,967,543 bonds are sold. This short selling is done to offset the \$19 million expenditure on one-year bonds with an equivalent value of four-year bonds.

\$19,047,619/0.7629 = 24,967,543

This transaction results in no net expenditure in the current year. In the next year, when the one-year bonds mature, the individual receives \$20 million, which can be used for their intended purpose. The four-year bonds, on the other hand, mature after three years, leaving the individual with an obligation to repay the counterparty \$24.968 million. This effectively locks in the loan at the present time.

Cash flows from this strategy (in millions of dollars):

Position	Year 0	Year 1	Year 2	Year 3	Year 4
Long 1-Year Bond	−19.048	20.000	0	0	0
Short 4-Year Bond	19.048	0	0	0	−24.968
Total	0	20.000	0	0	−24.968

$$\left(\frac{24{,}976{,}543}{20{,}000{,}000}\right)^{1/3} - 1 = 7.68\%$$

The interest rate can be calculated by dividing the amount to be repaid by the initial amount received, raising the result to the power of 1/3, and subtracting 1. In this case, the interest rate is calculated to be 7.68%. These forward rates are significant as they are not mere theoretical concepts but practical numbers that can be relied upon and used for transactions. This simple transaction in the open bond market allows individuals to obtain loans according to their preferred timing rather than being dependent on market conditions.

Coupon Bonds

The same principles that apply to pure discount bonds also apply to coupon bonds, with the key difference being the presence of intermediate payments in addition to the final principal payment. Coupon bonds can also trade at discounts or premiums to face value.

The standard Net Present Value (NPV) calculations are then applied, utilizing exchange rates across different dates. The rates are determined by the collective intelligence of the market participants. By summing up these rates, the value of the bond can be determined.

There is an important complexity to consider when quoting the prices of these bonds. Since different bonds have varying coupon rates, there is flexibility in selecting any desired coupon rate. For instance, a coupon rate of 1% of the face value or a high-yield bond with a 15% coupon rate can be chosen.

However, the significance of the coupon rate lies in its relevance to default risk and the preference for immediate payment. In the case of a default-free bond, the desired payment schedule can be specified, and the bond can be priced by discounting it at the appropriate interest rate.

Valuation of Coupon Bonds:

$$P_0 = \frac{C}{(1+R_1)} + \frac{C}{(1+R_1)(1+R_2)} + \cdots + \frac{C+F}{(1+R_1)\cdots(1+R_T)} \quad (4.5)$$

Bonds with higher yields generally have higher prices, while bonds with lower yields have lower prices. Nevertheless, the price of a bond accurately reflects the value of its cash flows in a fair market.

In terms of yield, it is common to quote the bond price not in relation to its coupon payments and NPV, but rather in terms of the bond's yield. This yield is calculated as a single number, denoted as "y", which equates the cash flows to the current market price. This yield is known as the yield-to-maturity (YTM) and is a complex average of all future spot rates.

The concept of yield will be further explored in the context of calculating rates of return for biotech investments, specifically venture capital. Venture capitalists commonly use a similar calculation known as the internal rate of return. Although this topic will be covered in more detail later, it is important to acknowledge its origin in bond pricing. The yield to maturity is the single number that summarizes the combination of coupon payments and principal value, resulting in the actual current price of the bond.

In practice, the market does not operate solely based on this simplified approach. Instead, various rates are employed to calculate the present value of coupons at different dates. However, to simplify bond valuation, it is common for individuals to quote bond values in terms of their yield rather than the specific price. For example, instead of stating a bond's price as $978.34, it may be referred to as a 3.5% yield bond or an 8.5% yield bond.

Corporate Bonds

The final subject to be discussed is corporate bonds, which involve borrowing from corporations and carrying a certain level of risk. To illustrate this, let's examine a development in the biopharmaceutical industry.

Axovant, a publicly traded company established with a focus on neurodegenerative diseases, including an Alzheimer's drug, secured a $55 million loan from a hedge fund venture debt investor. This funding was to support their drug development efforts, including a phase 3 trial for their Alzheimer's drug, Intepirdine, before it failed to pass key trials.

Axovant, a relatively young company, initially acquired a compound from GlaxoSmithKline and raised funds for its development. Notably, the last approved Alzheimer's drug was introduced in 2003. In addition to equity raised through their initial public offering (IPO), the company has turned to the bond market to raise an additional $55 million.

It is crucial to consider the risk associated with borrowing, particularly in the bond market. Rating agencies, such as Moody's, Standard & Poor's (S&P), and Fitch, play a vital role in assessing the risk of default for various bonds. These agencies assign grades based on the level of risk.

Table 6 - Credit ratings indicate the likelihood of default by each user

Credit Risk	Moody's	S&P	Fitch
Investment Grade			
Highest Quality	Aaa	AAA	AAA
High Quality (Very Strong)	Aa	AA	AA
Upper Medium Grade (Strong)	A	A	A
Medium Grade	Baa	BBB	BBB
Not Investment Grade			
Somewhat Speculative	Ba	BB	BB
Speculative	B	B	B
Highly Speculative	Caa	CCC	CCC
Most Speculative	Ca	CC	CC
Imminent Default	C	C	C
Default	D	D	D

These grades reflect the different risk categories for bonds. Prior to selling bonds to institutional investors like pension funds or mutual funds, they must undergo a rating process. While some bonds may trade without a rating, they tend to be smaller and less liquid.

In the case of Axovant's debt, since it was not a rated bond, it represented a private transaction between two parties. However, as a public company, Axovant was obligated to announce and provide details of the transaction.

On the other hand, when bonds are rated, the rating agency effectively endorses them after evaluating the financials and business model. This endorsement allows the bonds to be sold to a larger pool of investors.

It is important to note the distinction between the risk premium and the premium associated with corporate bonds compared to treasury bills. Academics have attempted to analyze and break down this difference.

The Promised Yield-To-Maturity (YTM) denotes the yield in the absence of default. The Anticipated YTM represents the average yield considering the probabilities of all potential outcomes. The Default premium signifies the disparity between the promised yield and the anticipated yield. The Risk premium (of a bond) is the distinction between the expected yield on a risky bond and the yield on a risk-free bond with comparable maturity and coupon rate.

In an attempt to analyze the subject matter comprehensively, the spread is categorized into multiple distinct classifications. These classifications include credit risk, default risk, liquidity risk, and market risk. It is crucial to approach this matter systematically by examining a bond and posing the inquiries about what it promises, what it is expected to deliver, and what disparities exist between those two aspects.

For instance, if a bond pledges to repay $1,000 but is currently trading at a price of $321.97, one can calculate the yield it will generate if it fulfills its promise. In this case, with a bond trading at $321.97 and promising $1,000 over a 10-year period, the promised yield amounts to 12%. However, the anticipated payment differs from the promised yield.

For the 10-year zero issued:

For the 10-year zero issued:

$$\text{Promised YTM} = \left(\frac{1000.00}{321.97}\right)^{1/10} - 1 = 12\%$$

$$\text{Expected YTM} = \left(\frac{762.22}{321.97}\right)^{1/10} - 1 = 9\%$$

$$\text{Default Premium} = \text{Promised YTM} - \text{Expected YTM} = 12\% - 9\% = 3\%$$

$$\text{Risk Premium} = \text{Expected YTM} - \text{Default} - \text{free YTM} = 9\% - 8\% = 1\%$$

A bond's default probability, determined by its rating, affects the likelihood of receiving the full $1,000. If, for example, the bond has a 70% probability of defaulting, one is more likely to receive $300 out of the $1,000. This expected payment leads to an expected yield of approximately 19% for this bond.

Over a 10-year period, it is expected that this bond will default around 24% of the time. Consequently, out of the $1,000, the investor can anticipate receiving approximately $760, resulting in an expected yield of roughly 9%. The disparity between the promised yield and the expected yield is known as the default premium, which represents the additional value received due to the bond's inherent risk.

Furthermore, the distinction between the expected yield and the risk-free yield, such as that of treasuries, is referred to as the risk premium. This risk premium is associated with market risk, liquidity risk, and other risks unrelated to default. Understanding these distinctions is of utmost importance when evaluating bonds. It enables investors to comprehend the sources of risk and determine whether they are comfortable managing these risks or if they align with their investment objectives.

To illustrate these differences graphically, the authors provide data from Moody's spanning almost a century, from 1920 to 2015. The data depicts the actual default rates of bonds with various ratings after different periods of time, ranging from one year to several years. Even after 20 years, a AAA bond, considered the safest and highest-rated bond, has a default rate of only 1.4%. This data emphasizes the significance of bond ratings, as they provide investors with transparency and reassurance based on historical analysis.

Conversely, bonds with speculative-grade ratings, often referred to as junk bonds, exhibit significantly higher default rates. After 20 years, over one-third of these bonds have defaulted. Consequently, junk bonds are considered highly risky and resemble equity more than traditional bonds. In the event of default, bondholders receive the collateral, which is typically the equity of the company, while equity holders face complete loss.

It is crucial to note that default rates fluctuate over time, particularly during different phases of the business cycle. Default rates during economic booms differ significantly from those during economic downturns. Therefore, it is essential to approach these risks with caution and acknowledge their complexity. The authors will delve further into these intricacies in subsequent discussions. Additionally, the recovery rates following a default vary considerably depending on whether the economy is experiencing a boom or a bust. The financial crisis, for instance, posed significant challenges for investors in terms of managing the recovery rates and loss rates associated with structured finance deals.

Financing Biotech through Bonds

The final point to be emphasized is the significant role that bonds have played, surpassing their previous impact. This development is highly stimulating. Drug royalty companies have been able to reinvest in various assets. These techniques are currently being employed, highlighting the importance of debt markets.

It is noteworthy to point out that biotech firms can involve the use of debt financing instead of equity financing for innovations. In other words, rather than relinquishing equity in exchange for funding, these companies can borrow money and repay it with interest.

This is particularly beneficial for entrepreneurs and some venture capitalists, as it offers non-dilutive financing, meaning they do not have to sacrifice a portion of the potential gains. The lenders' upside is limited to the predetermined interest rate.

For instance, Biopharma Credit regularly provides loans to biotech companies. If a company meets the necessary criteria, such as relatively low risk and a high likelihood of generating cash flows, it can secure loans to support its drug development projects. To illustrate the magnitude of these activities, Biopharma Credit conducted an initial public offering a couple of years ago, raising $762 million.

This capital is now being utilized to provide loans to various biotech companies, with previous funding amounting to approximately $2.5 billion. Several other organizations have also embraced the practice of lending money on a routine basis. Consequently, employing bonds as a means of financing drug development presents an intriguing alternative.

In the next chapters, we will demonstrate how to apply these tools to make capital budgeting decisions, calculate the cost of capital, and effectively utilize debt alongside equity to finance biomedical innovations.

5. THE CAPM & DIVERSIFICATION

In the previous chapters, the assumption was made that a specific value for the cost of capital was appropriate, taking into account the level of risk associated with the project. However, the explanation of how risk and the cost of capital are related had not yet been provided.

This chapter aims to address this gap by systematically examining the concept of risk and establishing a formal framework for estimating a project's opportunity cost of capital. The estimate is explicitly influenced by the level of risk associated with the project. To achieve this, the concept of diversification within a portfolio of assets is introduced.

The analysis heavily relies on uncertain future rates of return, which prompts a discussion on randomness and uncertainty. In this chapter, it is important to consider different scenarios from the perspectives of a biotech entrepreneur, a pharmacy executive officer, an investor, and a patient. By adopting the investor's perspective, one can better understand the opportunities and challenges faced by the industry.

Volatility, Risk, and Uncertainty

Volatility, risk, and uncertainty are three interrelated concepts. Volatility refers to the degree of variation or fluctuation in the price or value of a financial instrument over time. It is often used as a measure of market instability. Risk, on the other hand, is the probability of an adverse event occurring and the potential negative impact it may have on an investment. Uncertainty refers to the lack of predictability or knowledge about future outcomes. These factors play a crucial role in decision-making and portfolio management, as investors seek to balance potential returns with the associated risks and uncertainties. Let's delve into them.

Returns

The comparison between risk and reward is a fundamental concept that requires careful consideration, especially in the context of investing in biopharmaceutical assets. The valuation of assets is typically conducted through NPV analysis, which involves assessing the present value of revenues and deducting the present value of costs.

If the result is positive, the project is deemed viable; otherwise, it is not pursued. However, in practice, this process is far from straightforward, despite its conceptual simplicity. One of the main challenges lies in accurately estimating costs and revenues. While costs may appear to be easily determined, they can vary over time, especially in the case of biomedical RFID items.

Biomedical RFID items are electronic devices that use radio frequency identification technology to track and monitor medical equipment, supplies, and even patients within healthcare settings. These items are designed to enhance efficiency, improve patient safety, and streamline inventory management. By using RFID technology, healthcare professionals can easily locate and identify specific items, reducing the risk of errors and improving overall patient care. Additionally, biomedical RFID items can provide valuable data for analytics and decision-making, ultimately leading to better outcomes in the healthcare industry.

Moreover, unforeseen technologies may emerge, necessitating additional investments in plant equipment. Nonetheless, although challenging, this task is precisely what financial markets are designed for. Financial markets regularly evaluate and support various innovative technologies, even those that may seem unconventional at first, such as social media.

For instance, Facebook, a technology that few could have predicted would become a multi-billion-dollar company, has proven to be highly profitable. However, forecasting Facebook's revenue model two decades ago would have been extremely difficult. Financial markets and services enable us to address such challenges, as they are specifically designed for this purpose.

To tackle these challenges effectively, it is essential to measure and assess both risk and reward. Initially, the basic measure used is return on investment, a concept that is likely familiar to all readers. However, when it comes to biomedical projects, the definition of return on investment needs to be modified.

It is worth noting that throughout the book, readers are encouraged to adopt different perspectives, such as that of a biotech entrepreneur, a biopharma CEO, an investor, or a patient. These various viewpoints are essential for a comprehensive understanding of the subject matter, as we all play different roles at different times.

In order to address the challenges of the industry, it is necessary to consider different perspectives. It is suggested to adopt the viewpoint of a shareholder or investor in these technologies. To evaluate the performance, the rate of return, which compares the money invested to the money gained, is a useful measure.

By calculating the gross rate of return (cash out divided by cash in), one can assess the overall performance. The net rate of return is obtained by subtracting 1 from the gross rate of return, while the annualized net rate of return is determined by raising the gross rate of return to the power of 1/n and subtracting 1.

The cumulative return is a product of the gross rates of return. These metrics are important for shareholders and investors to focus on. It is necessary to calculate various characteristics of these returns, with a specific emphasis on risk and reward. Risk and reward are interconnected, as previously discussed in the introductory finance class.

Returns, being random variables, are unknown in the real world, and therefore need to be modeled using probability and statistics. Returns are often presented as a histogram or distribution, which may not be easily understood by most investors.

To cater to a wider audience, it is important to simplify and interpret the information. Various features of the distribution can be calculated, such as the mean, median, 75th percentile, and 5th percentile (representing losses). These measures provide insight into the risk and reward of investments. They help communicate with investors and provide a narrative of the investment opportunity. In finance theory, the expected return (average return) is commonly used to assess the reward.

However, this choice is not without its limitations, as will be discussed when examining multi-period investments using venture capital models and IRR. Calculating statistics for returns that have not yet been realized can be challenging. Despite these difficulties, the expected return is currently employed. The discussion on risk will continue in the next section.

Risk and Reward

Risk, as commonly understood, is assessed by the potential for negative outcomes. This is the prevailing notion when people refer to risk. One can say that investors hardly express concern about making excessive profits. The focus is always on potential losses. As a result, the left tail of the distribution is typically used to quantify this risk. However, there are instances where investors may experience financial losses when asset prices rise.

This occurs when engaging in short selling or certain options trading strategies, where a decline in asset value is desired. In such cases, relying solely on a downside measure of loss, represented by the left tail probability, may not be appropriate. Instead, it becomes necessary to consider the right tail of the distribution. To strike a balance, the approach is to examine variance and the spread of the distribution symmetrically, incorporating both the left and right sides. In the example provided, the narrower peak of the red curve indicates lower risk compared to the blue curve, which is more spread out and carries a higher probability of extreme returns.

It is worth noting that some investors may disregard one side of the distribution entirely, focusing solely on their preferred side. Certain hedge fund strategies, for instance, employ one-sided risk measures that solely consider the downside. However, these measures tend to be more complex and challenging to interpret.

As a result, the current approach favors an asymmetric measure that assesses the fatness of the distribution. This term is not derogatory but rather emphasizes that wider distributions with fatter tails are more likely to yield extreme observations. This can pose a problem if unexpected large losses occur, although some investors may be willing to accept these losses in pursuit of significant gains.

Ultimately, the choice of measure depends on individual preferences. Therefore, for the time being, a symmetric measure is employed. Returning to the initial class table depicting historical rates of return, it serves as a guide for understanding risk and reward. Notably, the data reveals a pattern where higher risk, as measured by standard deviation or volatility, generally corresponds to higher returns. Large company stocks, for instance, exhibit a volatility of 20% with an average return of 10% during the sample period from 1926 to 2015.

In contrast, small company stocks display a much higher volatility of 32%, yielding a 2% or 200 basis point increase in expected return compared to large stocks. Treasury bills, characterized by a low standard deviation of 3%, offer a corresponding modest expected return of 3%.

However, it is crucial to note that this risk-reward relationship is not an absolute truth based on theory or mathematics. While it holds true in most cases, there are exceptions where investors may determine that the risk premium is not worthwhile based on their risk tolerance and preferences.

In academic finance literature, the Sharpe ratio is a metric used to evaluate the excess expected return of an investment compared to the risk-free rate of return. It represents the risk premium divided by the risk. It is important to note that the Sharpe ratio does not consistently increase with different types of risks and can vary significantly across investments.

As a general guideline, investors typically prefer assets with higher Sharpe ratios and avoid those with lower ratios, although this is not always the case. The concept of a risk-reward trade-off is a graph that compares the volatility, measured using a rolling window of five years of daily data, to the average annualized return of the US stock market during the same period. According to financial theory, higher risk should correspond to higher returns, and vice versa. However, the graph reveals that during certain periods, higher risk is associated with lower average returns.

The graph showcases two lines: the red line represents the average annualized volatility, which was particularly high during the 1930s due to the Great Depression and the market crash of 1929. The blue line represents the average annualized five-year rolling window expected return. Surprisingly, despite the elevated volatility, the expected return was low. This inverse relationship between volatility and expected returns is contrary to what financial theory predicts.

One possible explanation for this phenomenon is the leverage effect, proposed by Fischer Black in 1972. According to this concept, when a company's value decreases, its equity becomes more highly leveraged, resulting in increased equity market volatility. However, this explanation does not fully account for the observed relationship between volatility and expected returns.

Oftentimes, investors react to high volatility by selling their equity holdings and moving their funds to cash. This mass selling further drives down equity market prices, leading to a negative relationship between volatility and expected returns.

These practical instances highlight the limitations of financial theory and the potential influence of behavioral patterns on market outcomes. While finance theory provides valuable insights and relationships, it does not always align with real-world observations.

Over extended time periods, there exists a positive correlation between risk and reward. When examining data spanning nearly a century, individuals generally act rationally. However, occasional instances of panic disrupt this relationship, necessitating caution. This is especially crucial for biopharmaceutical companies, particularly when seeking capital during market turmoil. Such periods are unfavorable for fundraising.

These circumstances pose significant challenges for companies planning their future strategies over five, ten, or twenty years. Conversely, if the company operates as an app developer, these concerns are less pressing. Their focus lies on short-term goals, such as increasing user base and app store performance.

On the other hand, if the objective is to develop a new drug for pancreatic cancer, planning often extends to a decade or more, making these effects highly significant. It is important to bear in mind that all the statistical findings we derive will have limitations. These limitations will be highlighted but it is first crucial to first analyze the functioning of the tools to comprehend their application before critiquing them.

Risk

With the aforementioned introduction, the subsequent analysis focuses on the concept of risk. Risk is volatility and can be illustrated using a series of graphs. The graphs depict the monthly returns of various assets, all plotted on a standardized scale ranging from -30% to 70%. Notably, these returns are not annualized but represent the actual monthly realized returns.

Treasury bills, which appear as a flatline due to the chosen scale. It is uncommon to observe returns exceeding 3% or 4% in a given month. Moving on to long-term treasury bonds and corporate bonds, corporate bonds exhibit lower volatility compared to government bonds. However, significant credit events associated with corporate bonds, such as the financial crisis, can lead to system-wide defaults. This introduces the notion that risks associated with corporate bonds extend beyond volatility and encompass gap risk.

Looking at the S&P 500, that represents the stock market, while fluctuations generally range between 10% and -10%, occasional severe declines are characteristic of stock markets. Investors factor in this possibility when making investment decisions.

Analyzing Pfizer, a prominent pharmaceutical company known for its stable cash flows. Despite this stability, Pfizer is considered risky when compared to the stock market. The volatility associated with investing in Pfizer and this highlights the potential for substantial gains or losses.

Next, reintroducing Genentech, a biotech company known for its extraordinary volatility, note that Genentech's trading ceased in December 2008 when it became part of Roche, but the volatility observed until that point was remarkable. Despite the risks, many investors profited from their investments in Genentech, while others incurred significant losses, particularly if they sold their holdings after a decline. This example illustrates the inherent risk and challenges associated with the biomedicine industry.

Understanding and properly measuring risk can enable its effective management. Furthermore, the traditional financing tools employed in the biopharma industry are inadequate for managing such risks. In the same way as using inappropriate tools to drive in a nail with a screwdriver, there is the need for more suitable financing instruments.

To further illustrate the distribution of returns, let's observe these histograms for Pfizer and Genentech. These histograms are overlaid with a red curve representing a normal distribution with the same mean and volatility as the actual data.

The red curve provides a reasonable approximation of the empirical distribution, except for the presence of fat tails on both ends. Despite the exceptional volatility associated with Genentech, the company's long-term performance appears impressive for those who held their investments from 1980 to 2009.

The upper tail exhibits a significantly large positive trend, whereas the left tail demonstrates a considerable magnitude. Moreover, the normal distribution does not align as effectively. This disparity is evident from the excessive thickness of the histogram and its heightened peak in the center.

Consequently, this presents a significant obstacle, particularly within the realm of publicly traded pharmaceutical and biotechnology companies. Additionally, when considering privately held companies, the associated risks are even more pronounced. Hence, it becomes imperative to strategize risk management approaches and, prior to that, ascertain methods for risk quantification.

The Five Levels of Risk & Uncertainty

In order to provide a conceptual framework, let's introduce a framework for understanding risk and uncertainty. This framework, developed in collaboration with physicist and proprietary trader Mark Mueller in 2010, consists of five distinct levels which are the relevant to drug development.

The first level is characterized by the absence of risk or uncertainty, where all information is known and certain. The second level involves risk without uncertainty, referring to statistical distributions that can be quantified. This is contrasted with the uncertainty described in the Ellsberg paradox discussed in a previous class.

The third level, termed fully irreducible uncertainty, suggests that with sufficient data and trials, uncertainty can be reduced to risk. Although the specific distribution may be unknown, a large sample size allows for reliable approximations.

The fourth level introduces partially reducible uncertainty, where a portion of randomness remains unknowable and incomprehensible. This is often referred to as "unknown unknowns" and poses significant challenges. Only a small number of investors are willing to navigate this type of uncertainty. The fifth level represents a limit to human knowledge, where attempts to understand are futile. It is typically advised to refrain from investing in anything characterized by zen uncertainty.

The progression through these levels can be explained by using the example of a solar eclipse. In ancient times, a solar eclipse was considered an uncertainty, as people lacked an understanding of its causes and predictability. The occurrence of a solar eclipse was viewed as a significant and potentially ominous event, leading to reactions such as the cessation of warfare.

In the present day, owing to the advancements in mathematics and physics, the state of affairs is entirely deterministic. It is worth noting that an upcoming solar eclipse is scheduled. It is possible that not all individuals are aware of this fact. For those interested in knowing the exact timing of a solar eclipse nowadays, NASA thoughtfully provides a countdown clock, specifying the remaining time in weeks, days, hours, minutes, and seconds. Such precision allows us to predict the exact moment of the eclipse occurrence with utmost certainty, down to the millisecond.

The disparity between the events that transpired in ancient times and those taking place today lies in our enhanced intellectual capabilities. We have acquired knowledge that enables us to transition between different domains. This is particularly relevant in the field of biomedicine, where we commence with a state of uncertainty and gradually progress towards risk.

If humans can successfully navigate this transition through scientific means, it would be highly advantageous. The swift approval of Herceptin by the FDA serves as an example. The drug did not undergo a mere trial-and-error approach, where chemicals were administered to patients haphazardly, hoping for positive outcomes. Instead, it was based on a solid theoretical foundation, a well-defined mechanism, and substantial evidence.

Numerous scientific papers were available, providing conclusive proof of the drug's efficacy. This transformation from uncertainty to risk instilled confidence in both investors and regulators. They recognized the substantial evidence and the ability to differentiate between patients who would benefit from the drug and those who would not.

Consequently, the FDA granted approval, leading to a remarkable response from patients who were not part of the clinical trial but sought admission due to the observed tumor shrinkage within a few days of Herceptin exposure. The strategy to address uncertainty involves minimizing it to a level of risk and, in exceptional cases, achieving complete certainty.

Returning to the realm of finance, while this approach may not resolve conflicts, it can certainly alleviate traffic congestion. Occasionally, we will have the opportunity to adjust these levels to various intermediate points.

It is important to highlight that we possess the capability to modify them not only scientifically but also financially. The aforementioned example demonstrates how scientific advancements facilitated the transition from level 5 uncertainty to level 1 certainty. In the upcoming weeks, we will explore how financial methods can be employed to transform uncertainty into specific types of risk.

Drug Development & Risk

The process of drug development aligns with the five levels of uncertainty. One can say that approved drugs generally offer a significant level of certainty for investors. For instance, Royalty Pharma recently acquired the royalty rights to Tysabri, a multiple sclerosis drug, for over $2 billion in cash. A substantial portion of this amount will be financed through investment-grade debt. This example exemplifies the level of certainty associated with approved drugs, attracting even risk-averse bondholders.

Phase 3 of drug development still entails some risk, as evidenced by notable failures in Alzheimer's phase-3 trials. However, this risk can be quantified to a certain extent. As the drug development process progresses to phase 2 and phase 1, the level of uncertainty increases.

The preclinical stage, where laboratory work and basic science experiments are conducted to understand cellular mechanisms and antibody interactions, is often considered to be irreducible or characterized by zen uncertainty. This high level of uncertainty makes it challenging for investors to predict returns and, consequently, discourages investment.

Understanding this taxonomy is crucial because it impacts investors. By examining the available funding for different projects, a similar inverted pyramid structure emerges, mirroring the clinical trials process. Projects with complete certainty attract a significant amount of funding, as demonstrated by the $39 trillion US bond market. Such investments can be securitized and easily purchased by bond investors.

The stock market, although smaller than the bond market, still garners funds despite its inherent risks. However, fully reducible uncertainty, where the distribution and calculations of mean and standard deviation are unknown, receives less funding. Venture capital investments, characterized by historical rates of return, also attract limited funding. Partially reducible uncertainty, associated with preclinical work, presents challenges in assessing the likelihood of any returns.

And finally, the funding of basic research is typically supported by philanthropists and government entities due to the absence of potential financial returns. This presents a challenge in securing funding, as investors tend to prioritize low-risk opportunities when considering different levels of uncertainty.

From a Sharpe ratio standpoint, investments with higher Sharpe ratios attract more capital, while those with lower ratios attract less. In our upcoming case studies and simulations, we will delve into the specific types of investors that are naturally inclined to invest in projects with varying levels of uncertainty. For instance, in situations of complete certainty, banks, money market funds, mutual funds, pension funds, and insurance companies are attracted to fixed-income securities as they seek to generate yield without taking on risk.

However, as we move down the inverted triangle, different types of investors become involved. To overcome this funding challenge, it is crucial to adopt a more systematic approach to identify potential investors and adjust the business model accordingly.

This would allow risk-averse individuals to focus on safer aspects of the project, while those willing to take on higher risks can handle the riskiest portions. In the field of biomedicine, the level of risk is increasing, but the human traditional response has been to avoid high-risk ventures and concentrate on low-risk ones. By utilizing more advanced financing tools, we can potentially address the funding cycles that occur in the industry.

The Capital Asset Pricing Model (CAPM)

This section delves into the capital asset pricing model (CAPM) by comparing systematic and idiosyncratic risk, describing a brief background on the CAPM, explaining the cost of capital in biopharma and concluding on the weighted average cost of capital (WACC).

Systematic vs. Idiosyncratic Risk

The capital asset pricing model CAPM pertains to a method of considering risk, specifically systematic risk versus idiosyncratic risk. Systematic risk refers to the risk associated with the overall movement of the entire market, while idiosyncratic risk represents a different type of risk with distinct implications for the cost of capital. From the perspective of an investor seeking to earn a cost of capital or a biotech entrepreneur aiming to generate that cost of capital from investors, it is essential to pose some fundamental inquiries.

When calculating the net present value (NPV) of investments using the cost of capital (r), it implies that only projects yielding at least r will be pursued. This ensures a positive NPV project, while projects earning less than r will be disregarded. Consequently, the value of r becomes crucial, as it represents the minimum rate of return necessary to satisfy investors. The cost of capital presents two primary challenges: the need for specialized knowledge in identifying risks and the requirement to manage those risks over time. Risk influences the cost of capital, but only certain types of risk. Hence an alternative explanation of this concept, focusing on systematic risk, will now be provided.

To illustrate, let's consider investing in two different companies: Rays Eyewear and Fresh Specs. The sunglasses company's revenue model is weather-dependent, losing \$100 per day on rainy days and earning \$200 per day on sunny days. Assuming a 50-50 chance of rain or shine, how much would one be willing to pay for this investment? The determination of the cost of capital can be inferred from the amount paid. For instance, if one pays \$1,000, it suggests a cost of capital of 5%.

The cost of capital is determined through an auction-like process, where the highest bidder establishes the cost of capital for the investment. Does anyone have an opinion on what they would pay for this investment? Perhaps around 10% seems reasonable, considering the inherent risk.

Similarly, let's examine another company, Raybrellas, which focuses on selling umbrellas. This company experiences the opposite cash flows, earning \$200 on rainy days and incurring a \$100 loss on sunny days, again with a 50-50 chance of rain or shine. How much would one pay for this company? The amount would likely be similar since it carries the same level of risk. If the cost of capital is 10%, one would presumably pay \$500 annually, assuming the \$50 figure represents the yearly amount.

Now, let's imagine the opportunity to allocate half of one's funds to each company. This results in a remarkable outcome: a return of \$50 regardless of rain or shine. This investment becomes perfectly certain. With a certain investment, the cost of capital will be significantly different. People would be willing to pay more for this investment due to the assurance of a definite return.

The essence of the capital asset pricing model is to eliminate the specific risk associated with individual assets by combining them into portfolios. By eliminating this risk, the cost of capital is altered. However, the risk that cannot be easily eliminated in this manner is the one that investors should be compensated for. If an investor cannot eliminate this risk, they will need to offer incentives to entice someone to take on that risk. In this case, there is no need for such incentives.

Even though the creation of a portfolio is not done explicitly for someone, the fact that they can do it themselves means that the collective wisdom of the market does not require any additional compensation for risk. This is because as long as there are entities that can create portfolios, such as a suntan lotion and a sunglass company merged with an umbrella company, they will provide a guaranteed rate of return.

In reality, perfect stories where assets are perfectly anti-correlated are rare. Most companies' earnings are positively correlated with each other. If all possible companies are included in a single portfolio, it will be the most diverse portfolio possible. After doing this, any remaining risk cannot be reduced through diversification. This remaining risk is known as systematic risk and is the risk that investors should be rewarded for when holding the market.

When calculating the cost of capital, regular risk is not a concern. The focus is specifically on systematic risk, which is what needs to be compensated. Unfortunately, idiosyncratic risk, which can be eliminated by combining assets like a sunglasses company and an umbrella company, will not be rewarded. If an investor chooses to take on this type of risk, they are responsible for it themselves.

This point can be illustrated through a simple example that demonstrates how only the idiosyncratic part of risk is eliminated when holding large portfolios, leaving the part that requires compensation. This will be further explained in the following section.

Brief Background on the CAPM

In 1964, the concept of the CAPM was introduced by Bill Sharpe, a finance professor at Stanford University. Sharpe's paper outlined the idea that the expected return on any asset is determined by the rate of return on risk-free investments, such as T-bills, along with a risk premium. This risk premium is a proportional bribe, reward, or sweetener that investors require to hold a portfolio of diversified assets known as the market portfolio.

The beta coefficient, a constant of proportionality, measures the association between an investment's return fluctuations and those of the market. If an investment moves in sync with the market, it will require the same risk premium as the market to attract investors. Conversely, if an investment moves counter-cyclically to the market, it can charge a premium to investors. The risk adjustment in CAPM is calculated by multiplying the market risk premium, historically around 7%, by the beta coefficient.

Mathematically, beta is determined by dividing the covariance between the asset and the market portfolio by the variance of the market. Linear regression can be used to estimate beta. The introduction of CAPM had a significant impact on the industry and became the foundation for mutual fund indexing, a multi-trillion-dollar business.

CAPM and its ex-ante version, which considers returns before they are realized, provide a risk decomposition framework for capital budgeting. The model enables investors to determine the risk premium required for an investment by separating it into two components: idiosyncratic risk, for which no reward is given, and systematic risk, which is correlated with the market portfolio and contributes to the risk premium.

To illustrate this concept, the example of window washers in Manhattan skyscrapers is often used. Despite the high risk associated with their job, window washers do not receive a significant risk premium due to the large number of individuals in New York City who are willing to work in various capacities.

The remuneration for high-rise window washers in New York City is a subject of great interest due to the associated risks involved in their line of work. Recent incidents have demonstrated the perilous nature of the job, as exemplified by an occurrence where two window washers were nearly involved in a catastrophic accident when one of the ropes they were using slipped. Although they were eventually rescued, their ordeal lasted for a considerable duration of time. Consequently, it is evident that this occupation demands utmost seriousness and caution.

To shed light on the salaries of these professionals, pertinent information can be found on a dedicated website that provides details about high-rise window washers in New York. Surprisingly, the average salary for an individual with moderate experience amounts to approximately $47,689, while those who are just starting out in the industry can expect to earn around $39,000.

On the other hand, individuals with extensive experience can command a higher salary of approximately $56,000. These figures may appear unexpectedly high, particularly to those considering summer internships in New York, where affordable housing is notoriously challenging to find. However, the remuneration is commensurate with the associated risks, adhering to the principle of risk-reward in the market.

To illustrate this concept further, consider a hypothetical scenario where a window washer incorporates tap dancing into their routine on the narrow catwalks. This form of entertainment would undoubtedly introduce additional risk. Consequently, one might assume that such window washers should be compensated more generously.

However, this assumption is incorrect, as the element of tap dancing is not an inherent requirement of the job. If a window washer chooses to engage in such activities voluntarily, they do so at their own discretion and will not receive additional compensation. Therefore, it is crucial to discern whether the risks associated with this profession are systematic or idiosyncratic, particularly from a pharmaceutical perspective.

Now to look at the concept of cost of capital in the field of biopharma by providing numerical examples.

The Cost of Capital in Biopharma

Investors in biotech companies have high expectations and seek substantial returns on their investments. Historically, the cost of capital in biotech has ranged from 20% to 40%, which is considered expensive from an entrepreneur's perspective.

Venture capitalists, in particular, aim for a much higher return on investment, typically requiring a company to increase its value five to tenfold. This is due to the high failure rate associated with biotech ventures. While the average return expected by venture capitalists is 20%, achieving such returns is challenging.

The historical internal rate of return (IRR) of biotech venture capital funds over the past few decades, illustrates a wide range of performance. Following a period of exceptional returns in the early 1990s and around the turn of the millennium, the subsequent decade witnessed underperformance, with returns falling below 20%. This was attributed to excessive investments in subpar projects.

However, recent years have shown improved returns, reigniting interest in biotech venture capital. Despite the risks involved, there is the necessity of high returns to compensate for the inherent uncertainties in the biotech industry.

Shifting the focus to pharmaceutical companies, let's compare their cost of capital to that of biotech companies. While pharma companies have historically had a cost of capital ranging from 8% to 15%, the recent period has seen a positive rate of return. Comparing these figures to the broader industry, it becomes apparent that the drug development business carries significant risk.

The phenomenon of whether a drug is effective or not is unrelated to the fluctuations of the stock market. This is due to the presence of scientific risk, which, by definition, is not correlated with market risk. One aspect is often overlooked when considering scientific risk: the necessity of funding a laboratory and the subsequent expenses of testing the drug on animals and humans through clinical trials that span several years.

The likelihood of securing funding for these endeavors is higher when the market is performing well compared to when it is performing poorly. Additionally, while evaluating scientific risk, it is important to consider financial risk as well.

Thus, a rolling window beta is employed to assess the pharmaceutical industry using an index. In this analysis, a regression is conducted using a 500-day period to determine the beta for the pharmaceutical sector. A beta value of 1 serves as the benchmark for the overall stock market. For most pharmaceutical companies, their beta is less than 1. This is because, regardless of market conditions, the demand for medications remains constant, making it a profitable business.

On the other hand, the beta for biotech companies is significantly higher due to the presence of financing risk in addition to scientific risk. Developing a therapy takes several years, and during a recession, it becomes challenging to secure the necessary funds for phase 2 clinical trials. This can lead to a situation where a company becomes financially trapped.

Biotech companies also face the risk of losing valuable human capital if they are unable to compensate employees, resulting in a decline in the company's value. Therefore, financing plays a crucial role in the biotech industry, as demonstrated by the cost of capital. This analysis focuses on individual companies rather than indexes and averages their betas.

It is evident that pharma companies have lower betas compared to biotech companies. Biotech companies are associated with high risk, a high cost of capital, and difficulties in obtaining financing unless systematic risk is minimized.

The CAPM is utilized to explore this relationship further. Assuming a risk-free rate of 4% and an expected return on the market portfolio of 10%, the risk premium is determined to be 6%. Beta values obtained from a website are used to calculate the cost of capital for pharma companies like Eli Lilly, which have a remarkably low beta of 0.48. This suggests that Eli Lilly's cost of capital for investments is 6.88%, a rate similar to that of a large mortgage with less than 20% down payment.

Conversely, biotech companies such as Agios, Alexion, Alnylam, and Editas exhibit higher betas. Editas, in particular, with a beta of 2.55, faces a risk premium 2.5 times that of the market. Consequently, when discounting future cash flows or assessing profitability, Editas employs a discount rate of 19.3%, which is comparable to a 20% rate. This analysis provides valuable insights into the interaction between finance and biomedicine, offering a convenient tool to estimate the value of a company.

Notably, Spark Therapeutics stands out with a beta of 1, which is surprising considering the risky nature of their gene therapy endeavors, such as injecting AAV vectors into children's eyes. However, this can be attributed to the fact that they have an approved drug priced at $850,000, or $425,000 per eye.

Due to the rarity of the disease, they treat and the significance of restoring sight to blind children, this price is considered reasonable, making it a lucrative business. Spark Therapeutics has effectively managed their business by engaging with payers in advance and informing them about the drug's progress in clinical trials, increasing the likelihood of approval.

Let's now turn to the idea of determining a mutually beneficial and fair price. Spark Therapeutics is distinguished from other companies by their understanding of the financial impact on payers and their desire to create a solution that benefits everyone. Investors are satisfied with a lower cost of capital of 11.68% compared to 20%.

To emphasize the importance of risk in various calculations, such as betas, let's use Solid Biosciences as an example. Despite a negative outcome in a clinical trial, the company attempted to increase the dosage, indicating their commitment to developing a therapy for a serious disease.

The company's stock returns are then volatile, with a beta higher than the market. The significant drop in the company's shares following a negative clinical trial outcome, illustrates the financial risks associated with investing in biotech. The ability to withstand such financial setbacks is crucial in the development of groundbreaking therapies.

Weighted Average Cost of Capital (WACC)

The Weighted Average Cost of Capital (WACC) is a financial metric used by companies to determine the average cost of financing their operations. It takes into account the cost of both debt and equity capital. WACC is calculated by weighting the cost of each source of capital by its proportion in the company's capital structure. This metric is important for businesses as it helps them evaluate investment opportunities and make informed decisions regarding capital allocation.

WACC is a relevant concept for large pharmaceutical companies with diverse projects. These companies often require a single hurdle rate for different projects, which is typically determined by the WACC. By understanding their WACC, companies can also assess their overall cost of capital and determine if it is feasible to pursue new projects or investments.

The calculation of the WACC is similar to the CAPM and portfolio theory. The value of a company is determined by the market value of its equity and debt. The rate of return on the assets is equal to the weighted average of the rate of return on equity and debt, based on the fraction of the company's financing from each source.

By calculating the beta of equity and debt, the beta of the assets can be determined, allowing for the computation of the cost of capital. While debt is often assumed to have zero beta, in certain cases, such as highly leveraged companies, it may have a significant beta.

Pharma and biotech companies use the WACC as a hurdle rate to assess the viability of projects. Pharma companies with lower betas have an advantage in taking on projects, while biotech companies face challenges due to their higher cost of capital.

Let's explore the application of portfolio theory in the next section, examining its potential benefits and limitations in the context of investing in compounds.

Portfolio Theory

Portfolio theory is a concept that aims to maximize returns while minimizing risk. It suggests that by diversifying investments across different asset classes, such as stocks, bonds, and real estate, investors can reduce the impact of any one investment's poor performance.

The theory also emphasizes the importance of considering the correlation between different assets when constructing a portfolio. By selecting assets that have low or negative correlations, investors can further reduce the overall risk of their portfolio. Overall, portfolio theory provides a framework for making informed investment decisions and managing risk effectively.

Brief Overview

The next question that arises is about the implications of systematic risk on portfolio theory. Interestingly, the development of finance theory took an opposite trajectory. Initially, the CAPM was employed to determine the cost of capital. However, the current approach involves examining the concepts of systematic and idiosyncratic risk to determine the construction of an optimal portfolio. Therefore, it is necessary to provide a brief overview of these concepts before applying them to the specific context of an orphan disease portfolio.

As described in the *simplifying financial terminologies* section of Chapter 1, in finance, a portfolio is defined as a collection of investments. The weights assigned to each investment, referred to as omegas, represent the dollar amount invested in relation to the total wealth.

$$\omega = \{\omega_1, \omega_2, \cdots, \omega_n\}$$

$$\omega_i = \frac{N_i P_i}{N_1 P_1 + \cdots + N_N P_n}$$

$$1 = \omega_1 + \omega_2 + \cdots + \omega_n \tag{5.1}$$

Essentially, the weights indicate the allocation of funds. Typically, these weights sum up to 1, as opposed to 0, the primary concern are non-arbitrage portfolios. The inclusion of arbitrage portfolios was specifically aimed at establishing the relationship between risk and reward.

It is important to note that these portfolio weights encompass most of the relevant information. However, as a drug developer or investor, there may be other aspects of a portfolio that are preferred to consider during deliberations. Nonetheless, for the assessment of risk and reward, the focus is chiefly on portfolio weights.

Asset	Shares	Price/Share	Dollar Investment	Portfolio Weight
A	200	$50	$10,000	10%
B	1,000	$60	$60,000	60%
C	750	$40	$30,000	30%
Total			$100,000	100%

To illustrate, let's consider an example where $100,000 sits in an account. Eventually, $10,000 is allocated to investment A, $60,000 to investment B, and $30,000 to investment C. The corresponding portfolio weights would then be 10%, 60%, and 30%, respectively. This translation of dollar amounts into fractions of total wealth invested demonstrates the concept of weighting. Now, to look at this using more sophisticated tools, specifically leverage.

In addition to purchasing stocks, a strategy can involve *shorting* certain types of assets, such as riskless bonds. Shorting a riskless bond entails issuing bonds and receiving cash from buyers, thereby borrowing funds. This is a significant situation due to the increased level of risk involved.

Specifically, the investment is made entirely in equities, totaling 200%. For instance, if an initial amount of $50,000 is leveraged to create a larger position of $100,000, the risk is heightened. This strategy is made possible by borrowing funds from external sources, which is a crucial aspect in meeting investor demands.

Asset	Shares	Price/Share	Dollar Investment	Portfolio Weight
A	200	$50	$10,000	10%
B	1,000	$60	$60,000	60%
C	750	$40	$30,000	30%
Riskless Bond	-$50,000	$1	-$50,000	-100%
Total			$50,000	100%

Some investors prefer high-risk investments, while others prefer low-risk options. By utilizing these tools, it becomes feasible to cater to the preferences of all investors while maintaining a reasonable rate of return for all parties involved.

The fundamental concept behind the portfolio is to move away from the traditional approach of selecting winners and losers. This notion can be challenging to shift in the minds of professionals within various industries.

For instance, venture capitalists in the biotech sector are specifically focused on selecting successful ventures. Therefore, proposing that they cease attempting to pick winners would be met with strong resistance.

This perspective contradicts the principles of portfolio theory, which aims to create diversified portfolios with a combination of strong, moderate, and weaker assets that collectively offer desirable properties. These properties primarily include meeting investor expectations in terms of expected return while minimizing risk. The goal is to construct a portfolio that offers the best risk-reward trade-off.

The Sharpe ratio serves as a useful metric in evaluating this trade-off. A high expected return relative to the risk-free rate, considering the level of risk taken, indicates that investors are being adequately compensated. Therefore, maximizing the Sharpe ratio becomes a key objective.

The question to now consider is how each asset contributes to this risk-reward trade-off. Does an asset enhance the overall portfolio's risk-reward trade-off? This is the primary application of portfolio theory, assessing the impact of asset combinations on the risk-reward trade-off.

Mean-Variance Analysis I

In the standard approach to addressing these matters, investors typically establish their preferences for risk and reward. The subsequent discussion will provide a concise overview of the diagram. The underlying concept revolves around acknowledging that investors hold preferences for different attributes.

As a result, the fundamental assumptions are as follows: individuals generally prefer more to less, experience diminishing marginal returns as the quantity of something increases, and exhibit transitivity in their preferences. However, it must be acknowledged that in practice, these assumptions may not always hold true due to the unpredictable nature of human behavior. Nevertheless, for the most part, individuals do tend to adhere to these basic assumptions.

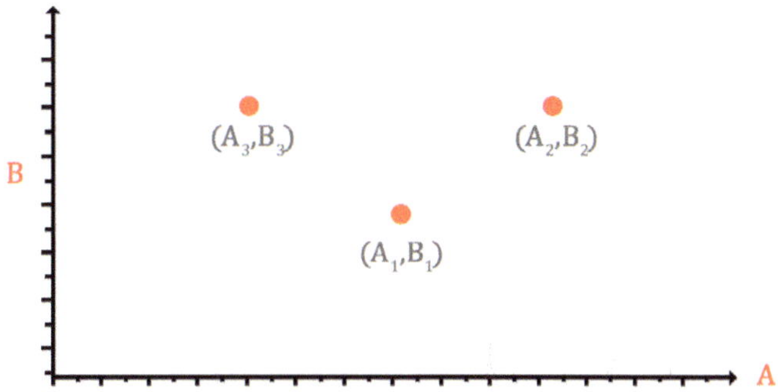

Figure 12: Rank orderings of bundles

When examining the various rank orderings of bundles, certain bundles cannot be easily compared. For instance, it is difficult to determine whether everyone would prefer over . However, it can be determined that everyone would prefer over .

This conclusion is based on the axioms that more is preferred to less and that an increase in both dimensions of A and B leads to an improvement in overall well-being. Of course, this assumes that the goods being considered are indeed beneficial and not harmful.

A good is defined as something positive, while a bad is defined as something negative. Therefore, if the items being evaluated are goods, it can be concluded that everyone would prefer over . The same reasoning applies to the preference for over , as has the same amount of B but less of A. Thus, due to the preference for more over less, a clear rank ordering can be established.

However, when comparing to , it becomes challenging to determine which option investors would prefer. This uncertainty arises because it depends on how individuals weigh the trade-off between A and B, making it difficult to discern a clear preference.

Consequently, the fundamental idea is to develop a method for ranking these various opportunities based on their characteristics. The two characteristics under consideration are 'goods' and 'bads'. The vertical axis represents the expected return, which corresponds to the level of excitement, happiness, and utility an individual experiences.

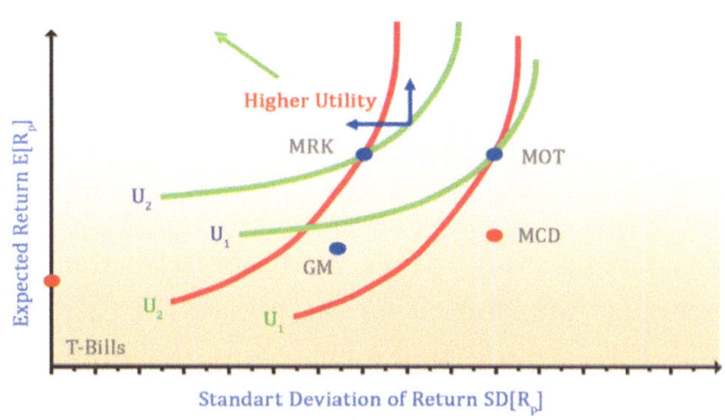

Figure 13: Expected Return and Standard Deviation of Return

A higher expected return leads to a greater level of satisfaction. On the horizontal axis, the standard deviation of the asset is measured, representing the level of risk. The larger the standard deviation, the less desirable the asset becomes, as individuals generally dislike risk and seek to minimize it.

The fundamental characteristics of portfolios and outlines can be seen as a derivation of the capital M, although the knowledge of the derivations is not necessary when using the capital M from capital budgeting. However, let's move on to specifically understanding how to apply these ideas to portfolio theory.

The underlying concept of portfolio theory is to construct a well-performing portfolio. Here, *well-performing* means that the portfolio should have a favorable expected rate of return and a desirable level of return volatility. Referring to Figure 13, the objective is to move towards the Northwest direction as much as possible. The aim is to construct a portfolio that is as Northwest-oriented as feasible in order to satisfy the investors.

$$\text{Mean} = \mathrm{E}[R_i] = \mu_i \tag{5.2}$$

$$\text{Variance} = \mathrm{Var}[R_i] = \mathrm{E}[(R_i - \mu_i)^2] = \sigma_i^2 \tag{5.3}$$

$$\text{Standard Deviation} = \sqrt{\mathrm{Var}[R_i]} = \sigma_i \tag{5.4}$$

$$R_p = \omega_1 R_1 + \omega_2 R_2 + \cdots + \omega_n R_n \tag{5.5}$$

$$\mathrm{E}[R_p] = \omega_1 \mu_1 + \omega_2 \mu_2 + \cdots + \omega_n \mu_n \tag{5.6}$$

$$= \mu_n \text{ (Weighted Average)} \tag{5.7}$$

To accomplish this, it is necessary to formulate the expression for the expected return of the portfolio, as well as the standard deviation of the portfolio. The expected return is calculated as a simple weighted average of the expected returns of the individual components of the portfolio. This is a straightforward and expected procedure, devoid of any surprises.

Mean-Variance Analysis II

Variance is a more intricate concept in finance, as it represents the expected squared return of a portfolio, taking into account deviations from the mean. The calculation involves determining the expected return minus the mean squared for the entire portfolio. The squared term is derived from multiplying the deviation by itself, which consists of multiple terms.

$$\begin{aligned}\mathrm{Var}[R_p] &= \mathrm{E}\left[(R_p - \mu_p)^2\right] \\ &= \mathrm{E}\left[\omega_1((R_1 - \mu_1) + \omega_2(R_2 - \mu_2) + \cdots + \omega_n(R_n - \mu_n))^2\right] \qquad (5.8)\end{aligned}$$

$$\begin{aligned}\mathrm{E}[\omega_i\omega_j(R_i - \mu_i)(R_j - \mu_j)] &= \omega_i\omega_j\mathrm{Cov}[R_i, R_j] \\ &= \omega_i\omega_j\sigma_{ij} \\ &= \omega_i\omega_j\sigma_{ij}\rho_{ij} \qquad (5.9)\end{aligned}$$

Specifically, there are n different terms in the deviation, resulting in n squared possible terms when multiplied by each other. This complexity arises from the inclusion of portfolio weights in the individual components of the portfolio. These n squared terms are determined by multiplying the portfolio weights by the covariances and expectations of cross-products of deviations.

Table 7 - Portfolio n-by-n Table

	$\omega_n(R_1 - \mu_1)$	$\omega_2(R_2 - \mu_2)$	$\cdots$	$\omega_n(R_n - \mu_n)$
$\omega_n(R_1 - \mu_1)$	$\omega_1^2\sigma_1^2$	$\omega_1\omega_2\sigma_{12}$	$\cdots$	$\omega_1\omega_n\sigma_{1n}$
$\omega_2(R_2 - \mu_2)$	$\omega_2\omega_1\sigma_{21}$	$\omega_2^2\sigma_2^2$	$\cdots$	$\omega_2\omega_n\sigma_{2n}$
$\cdots$	$\vdots$	$\vdots$	$\ddots$	$\vdots$
$\omega_n(R_n - \mu_n)$	$\omega_n\omega_1\sigma_{n1}$	$\omega_n\omega_2\sigma_{n2}$	$\cdots$	$\omega_n^2\sigma_n^2$

To calculate this, one can utilize an n-by-n table to compute all possible pairwise products and their expectations, with the total variance being the sum of these n squared terms. This calculation is crucial in assessing the risk associated with a portfolio.

It is important to note that portfolio risk is not solely determined by the risk of the underlying investments. Instead, it encompasses the risks of the individual investments as well as the potential covariances among different investments.

Portfolio managers, particularly in the biotech venture capital sector, are aware of this relationship and diversify their investments across different mechanisms to mitigate risk. By focusing on specific mechanisms and investing in a limited number of them, they avoid overinvesting in a single area and reduce the likelihood of correlated failures.

However, there is room for improvement in their approach, and the challenge lies in developing a more systematic framework for diversification. The section in this book on orphan diseases will delve into this topic, exploring the natural diversification that occurs in this context, which may not be present in other scenarios.

When describing the specific object of focus, the variability of the portfolio's return can be determined by the variance of the vector of weights multiplied by the vector of returns, similar to the approach used in the Arbitrage Pricing Theory (APT). The variance of the portfolio return can be expressed as a quadratic equation, where the weights are multiplied by all the possible entries of the covariance matrix and then multiplied by the weights again.

This alternative expression is particularly significant in financial analysis, as it represents the covariance matrix of all investments in the portfolio. The covariance matrix, denoted by the Greek letter sigma Σ, plays a central role in portfolio management, and considerable attention will be devoted to understanding different expressions for sigma, as well as estimating and controlling it.

The manipulation and comprehension of this matrix are essential in portfolio theory. The focus will be on managing the elements of the matrix, as they have a greater impact on portfolio outcomes than individual risks. The diagonal elements represent individual risks, while the off-diagonal elements represent the relationships between different risks.

Understanding and paying attention to these covariances are crucial in portfolio theory. The importance of covariances in portfolio management is emphasized by the fact that there are more off-diagonal elements than diagonal elements in the covariance matrix.

$$\begin{aligned}
R_p &= \omega_a R_a + \omega_b R_b \\
\mathrm{E}[R_p] &= \omega_a \mu_a + \omega_b \mu_b \\
\mathrm{Var}[R_p] &= \omega_a^2 \sigma_a^2 + \omega_b^2 \sigma_b^2 + 2\omega_a \omega_b \mathrm{Cov}[R_a, R_b] \\
&= \omega_a^2 \sigma_a^2 + \omega_b^2 \sigma_b^2 \qquad (5.10) \\
&\quad + 2\omega_a \omega_b \sigma_a \sigma_b \rho_{ab} \\
\text{Because } \rho_{ab} &\equiv \frac{\mathrm{Cov}[R_a, R_b]}{\sigma_a \sigma_b} \\
\mathrm{Cov}[R_a, R_b] &\equiv \sigma_a \sigma_b \rho_{ab}
\end{aligned}$$

To illustrate the concepts, a simplified two-asset example will be used, but the calculations can be generalized to any number of assets. The return of the portfolio is a weighted average of the returns of the two assets, with the weights summing up to 1.

The expected return is calculated by multiplying the weights with the mean returns of the assets. The variance of the portfolio, in this case, is determined by the weighted variances of the individual assets and the weighted covariance between the assets.

The covariance is expressed in terms of the correlation coefficient multiplied by the standard deviations of the assets. It is important to note that as the correlation between assets increases, so does the variance, indicating higher implicit risk. Investors seek investments that are negatively correlated or have low correlation to reduce portfolio risk.

The Efficient Frontier, Capital Market Line, and Tangency Portfolio

In the given example, a comparison is made between the financial performance of Pfizer and McDonald's from 2011 to 2016. Pfizer had an average monthly return of approximately 1% with a standard deviation of around 4.36% per month. On the other hand, McDonald's had a lower average return of 0.66% and a lower standard deviation of 3.73%. The correlation between the two stocks was found to be 44%. How would a portfolio of the two stocks perform?

$$\begin{aligned} \text{E}[R_p] &= \omega_{\text{PFE}}(0.0107) + \omega_{\text{MCD}}(0.0066) \\ \text{Var}[R_p] &= \omega_{\text{PFE}}^2(0.0436)^2 + \omega_{\text{MCD}}^2(0.0373)^2 + 2\omega_{\text{PFE}}\omega_{\text{MCD}}(0.440)(0.0436)(0.0373) \end{aligned}$$

The next step to solve this example involves analyzing the performance of a portfolio consisting of only these two stocks. The expected return of the portfolio can be calculated by taking a weighted average of the returns of Pfizer and McDonald's.

w_{PFE}	$E[R_p]$	$Var[R_p]$	$SD[R_p]$
0%	0.66%	0.002820	5.31%
10%	0.70%	0.002574	5.07%
20%	0.74%	0.002395	4.89%
30%	0.78%	0.002281	4.78%
40%	0.82%	0.002233	4.73%
50%	0.86%	0.002251	4.74%
60%	0.90%	0.002335	4.83%
70%	0.94%	0.002484	4.98%
80%	0.99%	0.002699	5.20%
90%	1.03%	0.002980	5.46%
100%	1.07%	0.003327	5.77%

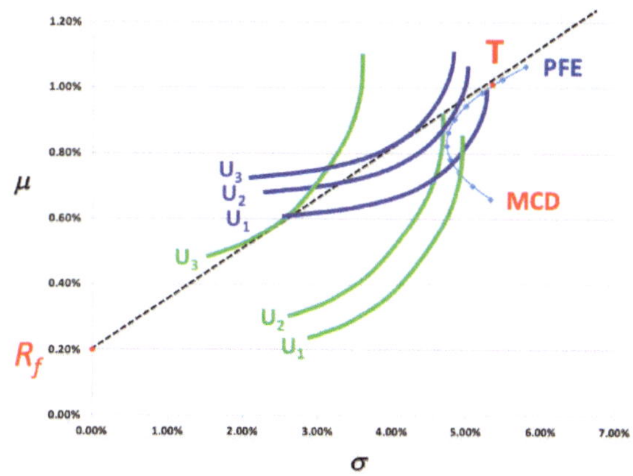

Figure 14: Risk-Return Tradeoff

Similarly, the risk of the portfolio can be determined by taking a weighted average of the individual stock risks, considering the covariance term and correlation. By varying the weight allocation between Pfizer and McDonald's, the risk-return tradeoff can be visualized.

Interestingly, it is observed that there exists an opportunity to improve the risk-return tradeoff in the middle range of weight allocations. By gradually adding Pfizer to a portfolio initially consisting of McDonald's, it is possible to decrease the risk and increase the expected return simultaneously. This outcome is highly desirable for investors as it allows for risk reduction and higher average returns.

However, the optimal portfolio allocation may vary depending on individual risk preferences. Different investors may choose different weight allocations based on their utility function and risk attitudes. Nonetheless, it is evident that holding only Pfizer or only McDonald's is not advisable.

The essence of portfolio theory lies in combining different assets to achieve a better risk-return tradeoff.

Furthermore, the introduction of a risk-free asset into the portfolio is explored. If a risk-free asset with a monthly return of 20 basis points (6% per year) is included, it is possible to further enhance the welfare of investors. By combining the risk-free asset with a specific combination of Pfizer and McDonald's, both investors can benefit.

The exact allocation of the risk-free asset will depend on individual risk preferences. The resulting portfolio positions of the investors will be more favorable than before, as represented by the northwest movement along the dotted line.

Returning to the exploration of a scenario where correlations undergo changes, the current correlation stands at 44%. This estimation is based on available data; however, let's assume the data provided is flawed and the correlation is not actually 44%. By employing the same calculation method, alternative risk-reward trade-offs can be determined by substituting a different value, such as 0.44.

Consider the following scenario: if a different value, say 0.25, is substituted, the correlation between Pfizer and McDonald's decreases, resulting in a shift from the blue curve to the red curve. In this case, it becomes crucial to ascertain whether investors are better off, worse off, or the outcome is undetermined. How can we establish that they are indeed better off?

By moving towards the northwest, specifically in the direction of the red curve, investors experience an improvement in their situation. The lower the correlation, the more advantageous it is for them. However, it is important to note that this does not imply we can predict investors' preferences.

To determine their desired position, their utility functions must be examined. Nonetheless, even without knowledge of their ultimate destination, investors are still better off due to the northwest movement.

Now, let's consider what happens when the correlation becomes even smaller, such as 0. In this scenario, the situation improves further. What if the correlation between Pfizer and McDonald's is negative? Well, in that case, the situation becomes even more favorable.

If the correlation reaches -100%, indicating perfect negative correlation, a spike emerges that intersects the vertical axis. This signifies that a combination of Pfizer and McDonald's exists, which carries no risk but offers a positive expected return.

Is it plausible to have an investment with no risk and a positive expected return that surpasses the risk-free rate? Why not? However, if such a situation were to occur, the expected return would decrease. For instance, if investors could obtain a riskless return of 85 basis points, they would seize the opportunity until the return diminishes to 0.2, which represents the equilibrium rate. Consequently, this scenario is unsustainable and not realistic. Nevertheless, it is included here to emphasize the significance of correlation and the search for negatively correlated assets.

Now, let's explore the opposite extreme. Suppose the correlation is 100% positive. In this case, the risk-reward trade-off is represented by a straight line. Combining these two securities offers no benefit as there are no savings to be gained. Consequently, it is more prudent to select the asset with higher yield and return, treating them as a single asset.

Finally, let's reintroduce the risk-free rate. When considering the risk-free rate alongside the standard situation, the trade-offs available to investors are as follows. However, when lower correlations are taken into account, the trade-off undergoes changes. This is evident in the trade-off with a 25% correlation or with a 0% correlation. Furthermore, as we move farther northwest, the slope of the line representing the trade-off increases. This slope is known as the Sharpe ratio.

Returning to the topic of the Sharpe ratio, if one aims to maximize this ratio, they are striving to move in a northwestern direction. This will result in improved outcomes for all investors. This is the primary concern in investment management, where the goal is to achieve the most favorable trade-off between risk and reward.

A concise way of summarizing this concept is by evaluating the Sharpe ratio. How far northwest can one go with their portfolios? This analysis applies not only to a two-asset scenario but also to a general case involving N assets.

In General:

$$\mathrm{E}[R_p] = \mu_a = \omega_1\mu_1 + \cdots + \omega_n\mu_n = \omega'\mu \quad (5.11)$$

$$\mathrm{Var}[R_p] = \sigma_p^2 = \Sigma_{i=1}^{n}\Sigma_{j=1}^{n}\omega_i\omega_j\mathrm{Cov}[R_i, R_j] = \omega'\Sigma\omega \quad (5.12)$$

Observations:

$E[R_p]$ is a weighted average of all the assets' expected returns

$SD[R_p]$ is smaller if assets' correlations are lower. It is less than a weighted average of the assets' standard deviations (unless perfect correlation)

The graph of portfolio mean/SD is nonlinear; find the "best" (quadratic optimization)

If we combine T-Bills with any risky stock, portfolios plot along a straight line

In essence, one must calculate the average of all the different investments and consider how the covariances and variances contribute to the overall risk. It is crucial to carefully manage these covariances by seeking assets with minimal correlation, although this is not an easy task. Once achieved, a combination of these assets will form the minimum variance boundary encompassing all the available assets.

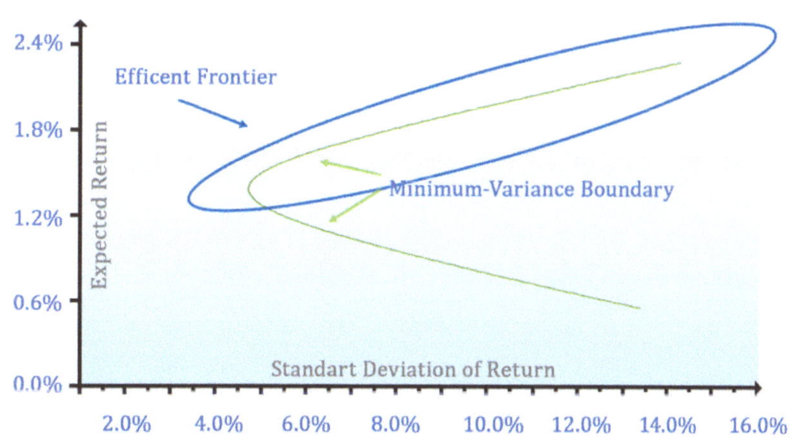

Figure 15: Minimum Variance Boundary and Efficient Frontier

The objective is to position oneself on the upper arc of the efficient frontier, which represents the optimal trade-off between expected return and risk. By focusing on this frontier, portfolio managers can effectively manage their investors' portfolios, striving to move as far northwest as possible.

If a risk-free asset is present, one can combine it with the tangency portfolio to guide investors along this line. It is worth noting that the curve generated here includes three assets: IBM, McDonald's, and Pfizer. None of these assets lie on the curve, indicating that it is never advisable to hold just one stock when multiple options are available.

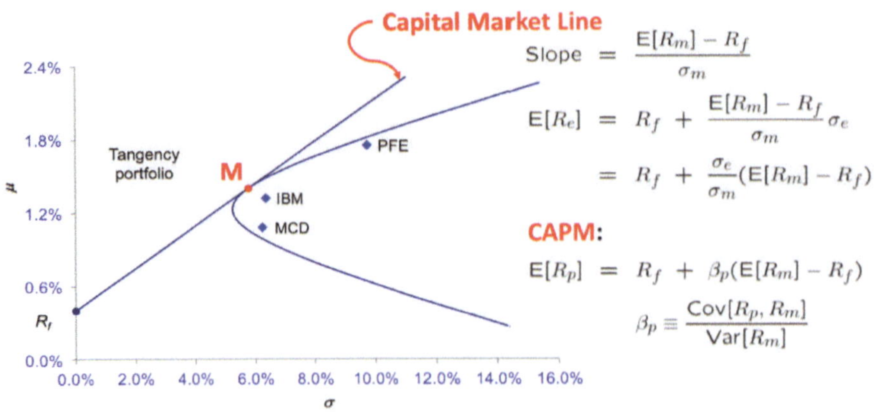

Figure 16: Capital Market Line

Capital Market Line:

$$\text{Slope} = \frac{\mathrm{E}[R_m] - R_f}{\sigma_m} \tag{5.13}$$

$$\begin{aligned} \mathrm{E}[R_e] &= R_f + \frac{\mathrm{E}[R_m] - R_f}{\sigma_m}\sigma_e \\ &= R_f + \frac{\sigma_e}{\sigma_m}(\mathrm{E}[R_m] - R_f) \end{aligned} \tag{5.14}$$

CAPM:

$$\mathrm{E}[R_p] = R_f + \beta_p(\mathrm{E}[R_m] - R_f) \tag{5.15}$$

$$\beta_p \equiv \frac{\mathrm{Cov}[R_p, R_m]}{\mathrm{Var}[R_m]} \tag{5.16}$$

This highlights the importance of diversification, as it is unwise to concentrate all investments in a single asset when there are others with lower correlation. This perspective contradicts the strategy of selecting winners and losers.

For those lacking such expertise, diversification offers a viable alternative. It is important to acknowledge the cultural gap between this approach and the mindset of drug developers, who tend to favor individual drug selection. In this class, we will have talented drug pickers and hunters, and it will be interesting to observe their reaction when we suggest the idea of constructing a portfolio of drugs.

They may express concern and emphasize the need for thoughtful decision-making and expertise. While these qualities are valuable, it is worth exploring alternative approaches for those without the same level of skill. As you already know, the tangency portfolio is synonymous with the market portfolio.

This concept brings us back to the distinction between market risk and idiosyncratic risk, aligning well with the arbitrage pricing theory and its notion of a portfolio with a unit beta relative to a common factor. This encapsulates the CAPM from the perspective of a portfolio manager aiming to construct the most well-rounded portfolio possible.

Special Scenario I

Now, let us examine a few special scenarios. The first case we will consider is the simplest one and explore what occurs when there are n positions of equal size across the assets, resulting in equal weighting across the n assets.

$$\begin{aligned} \text{Var}[R_p] &= \sum_{i=1}^{n} \frac{\sigma_i^2}{n^2} + \frac{1}{n^2} \sum_{i \neq j} \text{Cov}[R_i, R_j] \\ &= \frac{1}{n} \times \text{Average Covariance} + \frac{n-1}{n} \times \text{Average Covariance} \\ &\approx \text{Average Covariance} \qquad (5.18) \end{aligned}$$

This represents a valid portfolio, albeit not a particularly astute one. When the portfolio consists of equal weights for all assets, the portfolio weight is 1/n. Substituting this value into the formula yields a concise expression for the portfolio's variance. The variance is equal to the average variance of the individual securities divided by n, multiplied by the average covariance, which is then multiplied by .

As n approaches infinity, the first term vanishes, leaving only the risk of the portfolio, which is determined solely by the average covariance. The risk is independent of the risk associated with any individual investment and is instead influenced by the relationship between investments, emphasizing the importance of considering correlation.

Special Scenario II

The second special scenario illustrates the perspective of hedge fund managers and their enthusiasm for identifying assets with low correlation to existing portfolios. To explore this, further analysis is required. Specifically, we need to calculate the variance of a specific portfolio.

As previously shown, the variance of the portfolio's return is determined by the portfolio weight multiplied by the sigma matrix multiplied by the portfolio, resulting in a quadratic form. For this scenario, we will consider a portfolio consisting of n assets that exhibit the same pairwise correlation. In other words, any pair of assets will have the same correlation, denoted as ρ.

$$\Sigma = \sigma^2 \begin{bmatrix} 1 & \rho & \cdots & \rho \\ \rho & 1 & \cdots & \rho \\ \vdots & \vdots & \ddots & \vdots \\ \rho & \rho & \cdots & 1 \end{bmatrix} \quad , \quad \rho > -\frac{1}{n-1} \tag{5.19}$$

$$\text{Sharpe Ratio} = \frac{\mu}{\sigma} \frac{1}{\sqrt{\rho + (1-\rho)/n}} = \text{SR}\theta \tag{5.20}$$

This correlation matrix, known as the covariance matrix, is derived from the correlation matrix multiplied by the standard deviation's sigma squared, assuming identical standard deviations for all assets. However, it is important to note that the correlation must satisfy a specific inequality to ensure the covariance matrix is valid, as negative variances are not permissible.

$$\theta = \frac{1}{\sqrt{\rho + (1-\rho)/n}} \tag{5.21}$$

Where:

As n increases, θ also increases, with a limit of $1/\sqrt{\rho}$ when $\rho \neq 0$

If $\rho = 0$ then $\theta = \sqrt{n}$

Table 8 - Covariance Matrix

Correlation	Number of Assets n											
ρ	1	2	5	10	20	30	40	50	100	150	200	∞
90%	1.00	1.03	1.04	1.05	1.05	1.05	1.05	1.05	1.05	1.05	1.05	1.05
80%	1.00	1.05	1.09	1.10	1.11	1.11	1.11	1.12	1.12	1.12	1.12	1.12
70%	1.00	1.08	1.15	1.17	1.18	1.19	1.19	1.19	1.19	1.19	1.19	1.20
60%	1.00	1.12	1.21	1.25	1.27	1.28	1.28	1.28	1.29	1.29	1.29	1.29
50%	1.00	1.15	1.29	1.35	1.38	1.39	1.40	1.40	1.41	1.41	1.41	1.41
40%	1.00	1.20	1.39	1.47	1.52	1.54	1.55	1.56	1.57	1.57	1.58	1.58
30%	1.00	1.24	1.51	1.64	1.73	1.76	1.77	1.78	1.80	1.81	1.82	1.83
20%	1.00	1.29	1.67	1.89	2.04	2.10	2.13	2.15	2.19	2.21	2.21	2.24
10%	1.00	1.35	1.89	2.29	2.63	2.77	2.86	2.91	3.03	3.07	3.09	3.16
0%	**1.00**	**1.41**	**2.24**	**3.16**	**4.47**	**5.48**	**6.32**	**7.07**	**10.00**	**12.25**	**14.14**	**∞**
-10%	1.00	1.49	2.89	10.00								
-20%	1.00	1.58	5.00									
-30%	1.00	1.69										
-40%	1.00	1.83										
-50%	1.00	2.00										
-60%	1.00	2.24										
-70%	1.00	2.58										
-80%	1.00	3.16										
-90%	1.00	4.47										

Using the covariance matrix in Table 8, we can calculate the Sharpe Ratio for the portfolio. The Sharpe ratio has a straightforward expression, equal to the Sharpe ratio of each individual asset multiplied by a multiplier that depends solely on the number of assets and the correlation. For now, we will focus solely on this multiplier, disregarding the individual Sharpe ratios. The multiplier takes various forms depending on the values of ρ and n.

In the following example, an illustration is provided to explain the concept of asset correlation and its impact on the Sharpe ratio. It is stated that when assets have a high correlation, adding more assets to the portfolio does not significantly increase the multiplier or boost the Sharpe ratio.

However, when the correlation is lower, such as 40%, adding more assets can significantly increase the Sharpe ratio by up to 58%. It is emphasized that having assets with no correlation provides the greatest diversification potential and allows for unlimited multiplication of the Sharpe ratio. The importance of finding assets with no correlation is highlighted, as it can lead to substantial value creation and is often referred to as statistical arbitrage.

Portfolio Theory in a Biomedical Context

This section introduces mean variance analysis and its application to capital budgeting. Calculating the covariance matrix, denoted as sigma Σ, is necessary to construct portfolios. Historical data is commonly used to calculate sigma, but it is acknowledged that historical data may not always accurately reflect future conditions. Therefore, judgment is required to adjust the correlation values in the covariance matrix based on the underlying mechanisms and characteristics of the assets.

It is emphasized that incorporating scientific insights along with financial insights allows for forward-looking analysis and can provide valuable opportunities. The importance of combining these two perspectives is stressed to maximize the effectiveness of these analytical tools.

In a biomedical context, portfolio theory refers to the diversification of investments in healthcare projects or research initiatives to minimize risk and maximize returns. By allocating resources across a range of projects with varying levels of risk and potential rewards, portfolio theory enables decision-makers to make informed choices that can lead to improved outcomes in healthcare innovation.

This approach recognizes the importance of balancing investments in high-risk, high-reward projects with more stable and predictable ones, ultimately contributing to advancements in biomedical research and healthcare delivery.

Biomedical Portfolios

The implementation of portfolio theory presents challenges despite its basic principles being relatively straightforward. In the context of biomedical portfolios, the nature of assets differs from stocks, as they lack divisibility.

Unlike publicly traded companies, private companies in biomedicine cannot be purchased in fractional shares. Weights are discrete. For example, one cannot always have 1/3 of a project. Although it is possible to involve multiple investors to divide an asset into smaller parts, the level of divisibility is limited. This limitation often discourages companies from opting for an Initial Public Offering (IPO), as they prefer raising funds in smaller increments to facilitate investment digestion.

When venture capitalists invest in biomedical companies, they typically invest in the entire company rather than a fraction of it. This indivisibility of investment is a significant factor that must be considered when managing biomedical portfolios. Another important aspect is the binary nature of these investments, where they either succeed or fail.

This binary risk differs from the risks typically taken by most investors. For instance, if a farmer hires someone to plow their field and the job is only half-completed, the farmer still has half of their field plowed. However, in the case of biomedical investments, such as developing a drug, partial completion does not yield proportional value.

Denote Success/Failure of Project i by $I_i = \{0,1\}$

$$\text{Payoff} = (I_1 \times \text{rNPV}_1) + (I_2 \times \text{rNPV}_2) + \cdots + (I_n \times \text{rNPV}_n) \quad (5.22)$$

$$\Pr\left(\sum_{i=1}^{n} I_i = k\right) = \sum_{j=0}^{k} \binom{n}{j} p^j (1-p)^{n-j} \quad (5.23)$$

Due to the indivisibility and binary nature of biomedical investments, traditional tools of portfolio theory in modern finance are not directly applicable. This section aims to introduce some of the tools needed to demonstrate their practical application.

To account for the binary nature of investments, indicator variables are defined to represent success or failure i.e., a drug or treatment is either approved or rejected. Each investment's payout is determined by multiplying the risk-adjusted Net Present Value (NPV) with the corresponding indicator variable.

The properties of these indicator variables are crucial in understanding biomedical asset portfolios. Although not delving into extensive statistical analysis, a few examples are provided to illustrate the evaluation of a biomedical portfolio.

Incorporating correlation among different investments is another important consideration. To assess the diversification potential, it is necessary to determine the relationship between different drugs, such as drug A and drug B.

Figure 17: Normal Distribution

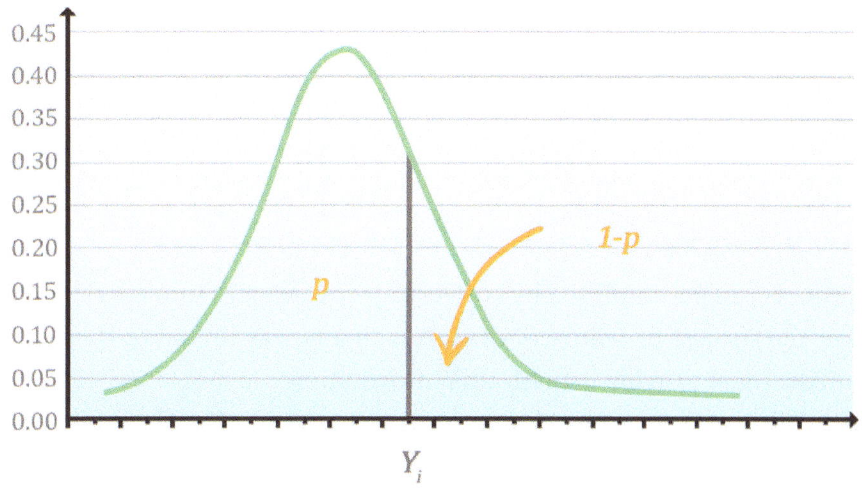

$$\Pr\left(\sum_{i=1}^{n} I_i = k\right) = \sum_{j=0}^{k} \binom{n}{j} p^j (1-p)^{n-j} \tag{5.23}$$

$$I_i = \begin{cases} 1 & \text{if } X_i > \gamma_i \\ 0 & \text{if } X_i \leq \gamma_i \end{cases}, \quad X_i \sim \mathcal{N}(\mu_i, \sigma_i^2) \quad \begin{bmatrix} X_1 \\ X_2 \\ \vdots \\ X_n \end{bmatrix} \sim \mathcal{N}(\mu, \Sigma) \tag{5.24}$$

This can be achieved by introducing latent variables, which are unobservable continuous variables denoted as . The success or failure of a drug is determined by comparing to a threshold value, . By assuming multivariate normal distribution and using a correlation matrix sigma, the correlation of success or failure among indicator variables can be modeled.

The process of incorporating correlation and modeling latent variables allows for a better understanding of biomedical portfolios. By assuming the existence of unobservable variables and utilizing data and calibration methods, these correlations can be estimated.

Challenges When Estimating the Correlation Matrix

The primary concern revolves around determining the value of sigma Σ.

How can the correlation between these latent variables be ascertained? This is where the situation becomes particularly intriguing. To comprehend the relationship among these distinct latent variables, it is necessary to consult with scientists. This is not merely a financial issue; it is a scientific problem.

$$\Sigma = \begin{bmatrix} \sigma_1^2 & \sigma_{12} & \cdots & \sigma_{1n} \\ \sigma_{21} & \sigma_2^2 & \cdots & \sigma_{2n} \\ \vdots & \vdots & \ddots & \vdots \\ \sigma_{n1} & \sigma_{n2} & \cdots & \sigma_n^2 \end{bmatrix}$$

$$= \begin{bmatrix} \sigma_1 & 0 & \cdots & 0 \\ 0 & \sigma_2 & \cdots & 0 \\ \vdots & \vdots & \ddots & \vdots \\ 0 & 0 & \cdots & \sigma_n \end{bmatrix} \begin{bmatrix} 1 & \rho_{12} & \cdots & \rho_{1n} \\ \rho_{21} & 1 & \cdots & \rho_{2n} \\ \vdots & \vdots & \ddots & \vdots \\ \rho_{n1} & \rho_{n2} & \cdots & 1 \end{bmatrix} \begin{bmatrix} \sigma_1 & 0 & \cdots & 0 \\ 0 & \sigma_2 & \cdots & 0 \\ \vdots & \vdots & \ddots & \vdots \\ 0 & 0 & \cdots & \sigma_n \end{bmatrix} \quad (5.25)$$

Ultimately, the process can be simplified by converting covariances into correlations. The question at hand is whether the success or failure of drug A is in any way linked to the success or failure of drug B. The scientists must provide us with this information.

Once we possess this data, it can be incorporated into the matrix. However, most scientists are unfamiliar with this particular setup. They have not engaged in modern portfolio theory, let alone latent variable estimation.

If one were to approach a scientist and inquire about the correlation ρ, they would likely be unsure how to respond. Therefore, it is necessary to translate scientific concepts into the statistical framework of portfolio theory.

The suggested approach is to engage in a conversation with a scientist and pose a series of questions.

As an example, consider that there are two drug candidates, A and B. The first candidate utilizes a VEGF inhibitor to treat breast cancer, which is an angiogenesis inhibitor that impedes blood development in tumors. VEGF inhibitors have been employed for a considerable period of time and operate through tyrosine kinase inhibition. Drug A treats breast cancer using a VEGF inhibitor, while drug B treats ovarian cancer with the same type of inhibitor.

Now, suppose drug A has failed. The question to the scientist is whether this failure impacts their opinion of drug B? The scientist's response is that the impact is minimal since drug A and drug B are distinct drugs designed to treat different diseases. However, both drugs inhibit VEGF.

Now, let's consider a second example. There are three drug candidates, C, D, and E. Drug C addresses gangliosidosis, specifically Tay-Sachs disease. Drug D treats mucolipidosis, and drug E utilizes gene therapy to treat Duchenne muscular dystrophy. Now, the question to the scientist is whether if drug C fails, their perspective on the prospects of drug E is altered.

The scientist's response is that the failure of drug C does not affect the prospects of drug E since they target completely different diseases with distinct causes. However, gangliosidosis and mucolipidosis are both lysosomal storage diseases that impact similar brain tissues. They would need to gather more information to determine the correlation between them.

In the first scenario, there is correlation due to the shared characteristic of solid tumors and the same mechanism of action, despite being different drugs targeting tyrosine kinase. Therefore, if drug A fails, careful consideration must be given to drug B. As for drugs C, D, and E, there are some similarities, but C and E are entirely unrelated.

On the other hand, both C and D are lysosomal storage diseases, necessitating further investigation. However, after thoroughly examining the mechanisms employed, which may differ significantly, a judgment can be made. This is where the role of judgment, guided by the framework, becomes crucial. Having consulted with a scientist and obtained their insights, the next step is to present the information in a format that can be effectively utilized by a financial economist aiming to optimize the portfolio.

While this process is not perfect and introduces some noise, it allows for a more systematic approach to considering diversification. Historical data is also utilized to estimate clinical trial success and failure rates, as well as correlations. However, in areas such as gene therapy and RNAi, where limited data points are available, scientific knowledge becomes crucial in determining these correlations.

Lastly, correlations are of lesser concern in rare diseases. Due to the distinct nature of rare diseases and the unique therapeutics developed for them, the failure of one therapy does not necessarily indicate the failure of others. This concept is exemplified by the orphan disease landscape, which serves as a clear illustration of how portfolio theory can be applied.

Applying Portfolio Theory to Rare Disease Portfolios

An orphan disease is defined as a condition that affects 200,000 patients or fewer in the United States, as stated in the 1983 Orphan Drug Act. This legislation aimed to prioritize the development of therapies for diseases with smaller patient populations.

Examples of orphan diseases include sickle cell anemia, estimated to affect around 100,000 patients in the US, and glioblastoma, a type of brain cancer that affects approximately 100,000 individuals. It is important to note that these numbers are approximate and subject to change. Other rare diseases to note are cystic fibrosis (30,000 patients), Lou Gehrig's disease (ALS) (30,000 patients), Gaucher syndrome (6,000 patients), hemophilia (20,000 patients), and Duchenne muscular dystrophy (7,000 patients).

There are also ultra-rare diseases, which may only affect a few hundred patients. One example is paroxysmal nocturnal hemoglobinuria (PNH), a blood disorder that affects a very small number of individuals, estimated to be around 300 to 400 in the US. PNH prevents the production of a specific surface protein in blood cells, leading to the immune system attacking these cells. Symptoms include discolored urine, fatigue, and various organ-related problems. Unfortunately, individuals with PNH typically do not survive beyond their 20s or 30s.

One of the challenges associated with dealing with rare diseases is the difficulty of diagnosing them. With such a small number of patients, local doctors may not possess sufficient knowledge about these diseases unless they are affiliated with renowned medical institutions. However, the situation has improved over the past decade, and diagnoses are now possible due to advancements in genetic sequencing.

It is worth noting that approximately 25 million Americans are affected by rare diseases collectively, surpassing the number of individuals with cancer. About 80% of these rare diseases are linked to genetic defects, either inherited or occurring during an individual's lifetime. Thanks to the Orphan Drug Act and subsequent developments, there are now 596 FDA-approved orphan drugs available for these diseases.

The terms *orphan diseases* and *rare diseases* are not interchangeable. Recent advancements in genetic biomarkers enable the stratification of patient populations into smaller groups of fewer than 200,000 individuals. This means that diseases previously considered prevalent, such as breast cancer, can now be identified in smaller subgroups based on specific genetic mutations.

For instance, Herceptin, a drug used to treat breast cancer in patients with certain genetic mutations, was approved as an orphan drug. This distinction is crucial because orphan drugs offer certain economic advantages, which will be discussed in the subsequent sections.

The orphan designation provides several economic benefits to pharmaceutical companies that propose orphan drugs. According to the Orphan Drug Act, these benefits include tax credits of up to 50% for clinical testing expenses, expedited review by the FDA, and a seven-year period of marketing exclusivity.

This exclusivity means that during those seven years, no other competitors can have their drugs reviewed by the FDA for the same orphan disease. In addition to the patent protection, this further safeguards the revenue stream for the drug company. Furthermore, for certain rare diseases, such as tropical, infectious, and pediatric rare diseases, the FDA offers a priority review voucher (PRV). This voucher allows the drug to be reviewed in eight months instead of the typical 12-month review period.

However, obtaining a PRV requires an additional payment of $2.7 million, which covers the costs of hiring additional staff to conduct the priority review. The PRV can also be sold to another party for any other drug application. Recent sales of PRVs have reached significant amounts, such as $125 million, providing additional financing for drug development.

The PRV program was not conceived by the FDA but was proposed by economists as a means to incentivize drug developers to focus on less popular diseases. The US congress passed the legislation allowing the sale of PRVs to provide drug developers with more funding opportunities for rare disease treatments.

Alexion Pharmaceuticals

The Orphan Drug Act of 1983 and other enacted legislations offer significant advantages to pharmaceutical companies that focus on addressing unmet medical needs. These benefits have had a tangible impact in the industry. An illustrative example of the profitability in this field is Alexion Pharmaceuticals, a company established in 1992 by Leonard Bell, a professor at Yale Medical School.

Initially, the company concentrated on developing a single drug known as eculizumab, marketed under the trade name Soliris. In 2007, the drug received approval for the treatment of

PNH, a rare disease affecting a small population of around 300 to 400 individuals in the United States and approximately 7,000 worldwide.

PNH is a blood disorder characterized by the immune system attacking red blood cells, resulting in symptoms such as anemia, fatigue, and bone marrow failure. Patients also face the risk of blood clots and strokes, making the disease both life-threatening and painful.

The average survival time after diagnosis ranges from 10 to 20 years, with patients requiring regular blood transfusions, typically two units per month. The disease significantly impacts the quality of life, causing progressive fatigue and other complications like thrombosis.

Soliris, developed by Alexion Pharmaceuticals, has proven effective in reducing or eliminating the need for blood transfusions in most cases. However, it does not alter the underlying disease progression or offer conclusive evidence of extending life expectancy.

The drug primarily improves the quality of life for patients, providing relief from fatigue, pain, and the burden of monthly transfusions. As a result, Soliris is considered a valuable treatment option for those suffering from PNH.

The development of Soliris took a considerable amount of time, with a 15-year period from the establishment of Alexion Pharmaceuticals in 1992 to the drug's approval in 2007. This prolonged duration of investment before generating revenue is uncommon in the biotech industry. Typically, companies develop drugs until a certain clinical phase, such as phase 2b, and then sell to big pharma, cash out, pay investors, pay the Venture Capitalists and then move on.

In the case of Bell, he persevered and pursued this concept from its inception to obtain approval and market the drug through the FDA. Consequently, the drug, known as Soliris, is sold to patients. Soliris has proven effective in reducing or even eliminating the need for blood transfusions. However, it does not necessarily alter the disease's specific progression.

There is insufficient compelling evidence to suggest that it enables patients to lead a normal life. Nevertheless, it appears to mitigate the occurrence of blood clots. It is primarily regarded as a drug that enhances the quality of life rather than a transformative therapy. It does not guarantee a complete and prolonged lifespan extension. Due to the drug's limited existence and a relatively small patient population, the extent of its life-extending properties remains unknown.

To acquire this knowledge, one can refer to the medical industry's literature, which offers a fascinating aspect. Numerous individuals are willing to participate in the study of different drug and disease aspects. This information is readily accessible online.

The Cochrane studies, conducted by a non-profit organization, provide comprehensive analyses and meta-studies that consolidate various research and literature. These studies offer insights into the current state of different therapies. In the case of eculizumab, a drug used in treating a rare disease called PNH, a 55-page report published in 2014 provides a detailed overview of the studies conducted, their outcomes, and the researchers involved. The report concludes that the prescription of eculizumab for PNH patients cannot be fully supported or rejected unless new evidence from a large-scale, high-quality trial alters this conclusion.

The trials conducted on eculizumab were relatively small due to the rarity of the disease and the limited number of patients. These trials faced criticism for being conducted by the drug manufacturer, which raises concerns about potential bias. Additionally, the trials lacked a significant sample size and adequate control groups, as desired by the researchers.

The report highlights these concerns openly, inviting readers to review the details within the 55-page report. The report concludes by urging caution in interpreting the trial results and suggesting the need for future trials on this issue.

However, the report does indicate that current evidence suggests eculizumab improves the health-related quality of life and reduces the need for blood transfusions compared to a placebo. It does provide tangible benefits to patients, who greatly appreciate the relief from fatigue, pain, and monthly transfusions associated with the disease. Nonetheless, the trials do not provide conclusive evidence regarding whether the drug saves lives.

Considering the cost of the drug, it becomes challenging to determine an appropriate price. While the majority of the cost may not be borne by the individual, it is worth noting the prices of other drugs such as Gleevec, which costs around $100,000 per year, and Harvoni and Sovaldi, which cost $84,000 per year for a cure.

In light of this, the price for the treatment in question is revealed to be $440,000 per year. Although this cost may seem exorbitant, it is important to acknowledge that the drug significantly improves the patients' well-being, even if it does not necessarily extend their lifespan.

In the broader context, it represents a relatively small number of patients. Therefore, from an insurance standpoint, is it truly of significance? Perhaps it is not a major concern. Perhaps it is acceptable. In fact, we are capable of enhancing their quality of life, although not necessarily extending their lifespan.

They genuinely feel that they can lead a more typical life, irrespective of its duration. Allow me to provide some insights into the operational framework of Alexion. They not only concentrate on this particular medication. This has been their primary focus for a period of 15 years. However, subsequent to that, they have developed two or three other medications, all within the domain of rare diseases, targeting different aspects of blood disorders and related problems.

Consequently, they have obtained approval for two or three additional medications. In terms of financial figures, in 2016, their earnings amounted to $3 billion, with sales totaling $13 billion, netting $400 million in income. The company's market capitalization is valued at $30 billion. A few years ago, they were even included in the S&P 500. They are presently a part of the S&P 500, representing this rare disease medication. According to the growth rate projected on finance.yahoo.com, which we considered during our analysis of cost of capital, Alexion's anticipated growth rate over the next five years is approximately 19%.

This signifies that it is one of the fastest growing and evidently one of the most esteemed companies on the US stock exchanges. It has received numerous awards, as evident from their official website. To illustrate how well investors have fared, since their initial public offering in 1996 (keep in mind that they were established in 1992), but prior to the approval of Soliris in

2007, their compound annual rate of return was approximately 21% per year from 1996 until last December.

This corresponds to a 61% standard deviation, indicating a considerable level of risk and fluctuation. However, the Sharpe ratio stands at 0.3. When comparing this to the stock market's Sharpe ratio of around 0.35, it becomes apparent why Alexion is so intriguing to investors.

Financing Orphan Drugs

There are numerous advantages associated with investing in rare diseases, which have experienced significant growth in orphan drugs since 1983. Prior to 1983, there were very few approved drugs for rare diseases. However, since then, the number of approved drugs has steadily increased each year. It is worth noting that six out of the top 10 grossing drugs across all diseases are rare disease drugs, indicating their financial profitability.

Although rare diseases often affect small patient populations, they are associated with high prices. The passage mentions the Rare Disease Act of 2002, which called for the involvement of the NIH in conducting research on rare diseases. In 2012, a specific division of the NIH, the National Center for Advancing Translational Sciences, was established to facilitate the translation of research ideas into clinical practice.

Recent data shows a significant spike in rare diseases. However, the question remains whether investing in rare disease companies is a wise decision. To answer this question, basic calculations are necessary. On average, the development of a rare disease drug project incurs approximately $75 million in costs.

The probability of successfully developing a rare disease drug is around 25%, based on industry statistics. Once successful, a rare disease drug can generate approximately $150 million in annual revenues for the remaining life of the patent, which is typically 13 years due to the 20-year patent life. This translates to a total of around $1 billion in revenue.

These figures are rough estimates and may vary depending on the specific drug indication pair. From a financial perspective, the expected rate of return is around 20%, which is considered favorable. It is possible to obtain exclusivity extensions for rare disease drugs, which can provide an additional six months of patent protection and other forms of exclusivity. However, this does not directly extend the patent life, although it has a similar effect.

The risk associated with investing in rare disease drugs is high, with a volatility of 56%. The Sharpe ratio, a measure of risk-adjusted return, for this investment is 0.28, which is relatively low compared to the Sharpe ratio of 0.42 for the S&P 500. This suggests that investing in the stock market may offer a higher risk-adjusted return compared to investing in rare disease drugs.

The numbers mentioned in this discussion are not arbitrary but are based on industry figures. To provide a benchmark, the Sharpe ratio for Spark Therapeutics, a company trading on the stock market since 2015, is around 0.28, similar to the previous estimate. However, it is important to note that these calculations are based on data up until March 8 and may vary if data from an earlier period, such as February 22, is used.

Roche acquired Spark Therapeutics for an approximate sum of $5 billion, resulting in a remarkable return of 120%. The nature of biomedical investment can be characterized as binary, wherein the outcomes are either highly rewarding or significantly unrewarding. Spark Therapeutics is an exemplary company, having introduced the first therapy approved by the FDA.

However, the rate of return on their stock from 2015 until February 22nd was a mere 0.8% per year on an annualized basis, indicating an annual earning of less than 1%. Subsequently, the rate increased to 22% in a single day, although it should be noted that these figures do not consider the irregularity in returns.

$$\begin{aligned}
\text{Total number of projects} &= m \\
\text{Cost per project} &= C \\
\text{Cost of capital} &= R \\
\text{Duration of clinical trials} &= q \\
\text{Earnings per year if approved} &= X \\
\text{Payoff per project on arrival} &= Z \;=\; \frac{X}{R}\left(1 - \frac{1}{(1-R)^{20-q}}\right) \\
\mathrm{E[NPV]} &= \sum_{k=0}^{n} \mathrm{Prob}(H = k)(kZ) \\
\mathrm{E}[R] &= \frac{\mathrm{E[NPV]}}{mC} - 1 \\
\text{Annualized } \mathrm{E}[R^a] &= \left(\frac{\mathrm{E[NPV]}}{mC}\right)^{1/q} - 1
\end{aligned}$$

In light of these circumstances, the concept of portfolio theory becomes relevant. Instead of investing solely in rare disease companies like Spark Therapeutics, it is advisable to diversify investments across multiple such companies. Let us consider investing in of these companies, allocating dollars for each.

The cost of capital is denoted as , and the clinical trials require years to be completed. If successful, the revenues generated from years 8 to 20 amount to dollars per year, and the amortization is calculated accordingly. The ultimate payoff is determined by multiplying the indicator variables of various rare disease companies by the notion of a billion-dollar success.

By employing this approach, the relevant figures can be computed. Although the complexity of the calculations may seem overwhelming at first glance, prolonged contemplation reveals the inherent beauty in the process, albeit not necessarily indicative of its desirability. The Sharpe ratio serves as a measure of the risk-adjusted return obtained through these investment strategies.

By way of comparison, Warren Buffett, widely regarded as the most successful investor in public markets, possesses a Sharpe ratio of 0.75. Remarkably, with a portfolio consisting of twenty rare disease companies, the textbook's approach outperforms Warren Buffett's strategy. It is important to note, however, that this analysis is an abstract representation and does not incorporate all relevant variables. Therefore, caution should be exercised before directly applying these findings.

Divide and Conquer

In the realm of financial approaches to investing in various diseases, there is an intriguing method known as the orphan drug model. This model serves as a template for drug development, whereby the focus is on providing therapy to a specific group of patients.

Over the past few decades, there has been a shift towards a more sophisticated perspective, involving the stratification of patient populations. The identification of subpopulations, such as those with the HER2 gene who respond to Herceptin, has become possible. Furthermore, this stratification can be further subdivided, encompassing various patient populations.

Consequently, the concept of targeted therapy has become a reality. Instead of developing treatments for a large and diverse patient pool, the focus has shifted towards subsets. This approach aligns with the financing of these opportunities, as it allows for the division and mitigation of risks. By stratifying risks, investors can significantly reduce their exposure and attract more funding to the field.

This appears to be the direction in which the industry is heading. However, the question of whether it is feasible to address all diseases through this approach can only be answered by biomedical experts. Additionally, there may be an optimal level of stratification, as developing therapies tailored to each individual would not be financially viable.

While it may be possible to create a $5-million therapy for a single person, the cost-effectiveness of such an approach is questionable. The essence of this industry lies in the ability to develop therapies that can be applied to a large number of people, thus leveraging its impact.

The orphan drug model emphasizes the need for a balance between stratifying patients to some extent and avoiding excessive individualization, as this would undermine the economic model. It can be likened to insurance, which operates effectively when there is uncertainty about who will require coverage. However, if it is known with certainty that certain individuals will fall ill, it becomes a financing issue rather than an insurance problem. Similarly, the process of stratification must consider pricing factors.

Pricing

In order to gain a comprehensive understanding of the connection between pricing and financing, it is crucial to delve into pricing considerations. Determining appropriate pricing for drugs and the optimal level of differentiation is essential for assessing whether investors can expect a reasonable return on investment. Consequently, the question arises: what price should these drugs be set at?

In the realm of pharmaceutical pricing, the cost of a certain medication has skyrocketed from a few hundred dollars to a staggering $89,000. The discounted price remains uncertain, but the disparity is substantial. Furthermore, the situation becomes more intricate when we consider the actions taken by the certain company responsible for this change.

One must question the value of obtaining approval for a drug's specific use, which had not been sought previously. It appears that the company did not assume much risk, as the drug's safety had already been established through approval in other countries for different purposes.

Clinical trials were also not as extensive as one might expect, as the company relied on data from previous studies conducted 15 to 20 years ago, which they purchased and had not been analyzed. Thus, the number of studies conducted by the company was relatively small, involving only around 10 to 30 patients.

The company's approach can be seen as a financial arbitrage, capitalizing on the fact that safety concerns were already addressed in other countries, although not in the United States. They identified the drug as a treatment for a rare disease affecting a small population, specifically children, and obtained aPRV).

Additionally, the company did not bear significant risk in conducting clinical trials due to the limited number of patients involved. Consequently, they were able to assemble the entire package successfully. Congress is now demanding that the company disclose their research and development expenditures, although this information is not yet available.

The FDA granted the company expedited review, providing them with numerous advantages and a seven-year protection against competition. Furthermore, the FDA awarded them a PRV. When asked why the FDA extended such favorable treatment, the agency responded by stating that the drug in question had never been approved for any use in the United States.

Under US law, it was evaluated as a new drug, assessing its safety and effectiveness for the specific conditions listed in the labeling. Although versions of the drug are available in other countries for different indications, none were approved for the treatment of Duchenne muscular dystrophy (DMD) until now.

The FDA-approved labeling includes safety and clinical information specifying the drug's use for DMD. If a drug meets the necessary requirements for orphan drug designation, expedited programs, and rare pediatric disease designation, the sponsor becomes eligible for the associated benefits. In conclusion, the FDA's role in this matter is limited.

The pharmaceutical company in question has adeptly exploited the system, benefiting their investors greatly. However, Congress is now poised to address whether it is justifiable to charge $89,000 for a drug that was available for a mere $1,000 annually just a few days ago.

This complex situation arises from the actions taken by a certain party. It is distinct from a typical case of price gouging, where a drug already approved in the US experiences a drastic price increase. The moral intricacies at play here are understandable, as it is not simply a matter of celebrating capitalism.

The party in question is indeed making significant contributions to patients suffering from

DMD, which is commendable. During a conversation with a patient's father, an impressive statement was made. The father expressed support for capitalism, acknowledging the value of having a drug manufactured in the US with FDA approval and adherence to stringent production processes ensuring safety, efficacy, and purity. While he is willing to pay around $10,000 annually for the drug, the exorbitant price of $89,000 is deemed outrageous.

This goes beyond economic considerations and delves into the realm of ethics. The pricing matter cannot be disregarded, and while there is no definitive solution, it is important to contemplate this issue as it will resurface in the future. The situation is currently evolving, with the party in question retracting their pricing proposal.

Congress has requested information regarding research and development costs, prompting the need for a new pricing proposal if they wish to market the drug. It will be intriguing to observe the outcome, and it is hoped that in the coming weeks, further insights will be gained to inform our perspective on this matter, as a resolution is currently lacking.

6. OPTIONS

An option is a financial instrument that grants an individual the right, though not the obligation, to participate in a financial transaction. A call option confers the right to purchase a stock on or before a specified exercise date, while a put option grants the right to sell a stock within the same timeframe. The historical significance of options is multifaceted. Notably, the publication of the renowned Black-Scholes-Merton paper on option pricing in May 1973, co-authored by Black, Scholes, and Merton, holds great importance.

This academic work, which presented a rational theory for pricing options, coincided fortuitously with the establishment of the Chicago Board Options Exchange (CBOE) in April 1973. The CBOE facilitated the trading of exchange-traded options, standardized contracts accessible to all traders. The simultaneous emergence of the CBOE and the mathematical formula for pricing options pioneered by Black, Scholes, and Merton was an exceptionally fortunate occurrence.

Their formula subsequently became the industry standard for pricing various financial securities, leading the Chicago Board Options Exchange to officially adopt it as the option pricing formula in 1975. It is crucial to note, however, that the Black-Scholes-Merton formula is not the sole method for pricing options. Deviations from their model are frequently observed in option prices. Nevertheless, the framework they introduced remains highly influential.

Coinciding with these developments, Texas Instruments introduced the SR52 calculator, a revolutionary device that enabled users to perform complex trigonometric and algebraic computations with ease. This technological advancement further facilitated the adoption of the Black-Scholes-Merton formula.

Additionally, an intriguing aspect of this narrative emerged recently: the prevalence of Texas Instruments calculators among floor traders at the Chicago Board Options Exchange was due to the efforts of Irwin Guttag. Guttag, a founding member of the exchange and a member of its executive committee, influenced the adoption of these calculators as standard tools for traders. Remarkably, this decision was influenced by Guttag's mathematically proficient son, who, as a high school student at the time, demonstrated remarkable aptitude with computing instruments that were available during that era.

In the realm of finance, the emergence of new calculators sparked a significant development. Irwin's son, John, diligently studied the instruction manual and became proficient in programming the device. John successfully programmed the Black-Scholes formula, which he then presented to the floor traders. Consequently, the calculator quickly gained popularity among all floor traders, contributing to the widespread adoption of the Black-Scholes formula. This anecdote holds interest as John Guttag, the son of Irwin, is a well-known figure who previously served as the chairman of the electrical engineering and computer science department and continues to be a faculty member at this institution. This connection between computing and finance carries considerable significance.

However, it is worth noting that this connection did not come without a substantial cost. At the time, the calculator carried a price tag of $400, equivalent to approximately $1,800 in today's currency. Nevertheless, from the perspective of the traders, this cost was insignificant compared to the benefits it provided in terms of rapidly determining the value of call and put options. As a result, the options market experienced exponential growth, as depicted by the annual US options trading volume across various exchanges. This growth pattern mirrors the typical hockey stick graph seen in most financial securities, with the market expanding significantly after the introduction of the Black-Scholes formula in the 1970s.

A noteworthy milestone occurred in 2000 with the establishment of the International Securities Exchange (ISE). This electronic exchange, founded by Bill Porter, exclusively facilitated options trading. The success of ISE played a pivotal role in the creation of the building we are currently in. Additionally, it is important to acknowledge the existence of non-exchange-traded options, known as over-the-counter (OTC) options. These derivative securities are traded directly between two parties without the involvement of an organized exchange. The OTC derivatives market surpasses the size of the world GDP, reaching trillions of dollars. This remarkable scale is a testament to the ability of these securities to facilitate risk transfer to those best equipped to manage it, while relieving those who are ill-suited to bear such risks.

In recognition of their groundbreaking work on option pricing, Bob Merton and Myron Scholes were awarded the Nobel Prize in 1997. It is worth noting that Fischer Black would have also received this prestigious honor, had he not passed away from cancer several years earlier.

When utilizing discounted cash flow analysis to assess the value of a project, the potential for project expansion, reduction, or abandonment by managers should not be disregarded if circumstances do not unfold as anticipated. Projects that can be adjusted in response to new information are inherently more valuable than those lacking such flexibility.

Moreover, the value of this flexibility increases in tandem with the level of uncertainty. This adaptability bears resemblance to a financial instrument known as an option, which grants the holder the right, but not the obligation, to buy or sell a stock at a predetermined price on or before a specified date.

In this chapter, the formal connection between financial options and project flexibility will be expounded upon. The chapter commences with an examination of financial option theory, followed by an exploration of a closely related concept called a decision tree. Decision trees are commonly utilized to outline the available choices for project managers, identify project risks, and determine the impact of decisions on future cash flows. These structures are straightforward and are frequently employed to illustrate the real options embedded within capital investment projects.

Lastly, this chapter introduces the Monte Carlo simulation, a tool used to evaluate more intricate projects that cannot be analyzed using decision trees. Although the examples provided in this chapter pertain to the life sciences, it is crucial to note that the concepts and techniques discussed are applicable to numerous other industries and applications.

Option Pricing Theory

An option is a type of derivative security that grants the holder the right to engage in a financial transaction. Its value is derived from another security, and it comes in two basic forms: calls and puts. There are also variations such as American versus European options, with the former being exercisable before the maturity date and the latter only on the maturity date.

Other parameters include the strike price and exercise price. Interestingly, options are similar to insurance contracts, with put options serving as insurance on the value of a stock. The cost of the option is the deductible, and it shares many similarities with typical insurance contracts.

The distinctions between an option and an insurance contract are as follows: Firstly, an option can be utilized before its maturity, whereas an insurance contract can only be utilized after an accident occurs. Secondly, an option can be sold if desired, whereas auto insurance cannot be transferred. It is non-transferable and only applicable to the individual and their vehicle. Lastly, options are safeguarded against dividend payments. When the underlying asset's value decreases due to dividend payments, the exercise price of the option is adjusted. Conversely, an insurance contract does not automatically adjust for incidents.

The payoff of a call option is intriguing because it is a non-linear instrument. This means that the value of the call option's payoff is positive only when the stock's underlying value exceeds the strike price.

Terms:

- Strike Price K
- Time to Maturity $\tau = T - t$
- S_t = Stock Price
- K = Strike Price
- C_t = Call Price
- P_t = Put Price

$$C_T = \text{Max}\,[\,0\,,\ S_T - K\,]$$

$$P_T = \text{Max}\,[\,0\,,\ K - S_T]$$

At Maturity T, payoff

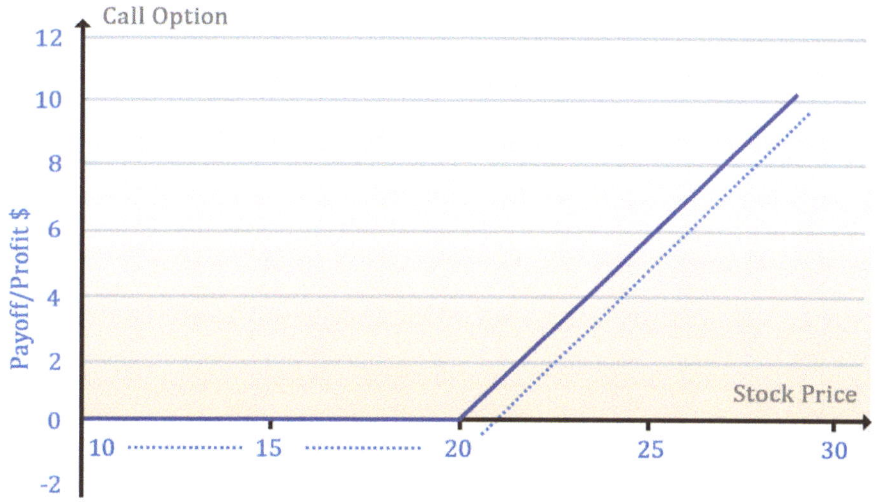

Figure 18: Call Option

If the stock's value falls below the strike price, the option loses its value as it grants the right to purchase the stock at a price higher than its current value. Therefore, if the stock is trading at $15 and the option allows purchase at $20, it holds no value and is represented as 0.

However, the current worth of an option differs from its payoff as it takes into account the time remaining until the option matures. Even if the stock is currently trading at $15 and the option has a payout at $20 and above, it can still hold value if there are several days left before it expires.

Within that time frame, the stock's value could potentially increase by $5 or $10, or even more in the case of a biomedical stock with a clinical trial readout. Therefore, the value of an option today versus its payout at maturity is influenced by the possibilities during this interim period.

When assessing the value of an option, it is sometimes examined on the payout day, resulting in either 0 or a linear function above the strike price. However, to determine its present value based on the stock price, the cost of the option must be subtracted. Valuing the option requires a thorough analysis, including the determination of its cost.

Once the cost is established, the net payout, which subtracts the cost, is represented by a dotted line. Consequently, for a range of values, the investor experiences a loss due to the upfront payment for the option. Unless the stock generates a certain amount of profit beyond the strike price, the net investment remains negative.

The following is a depiction of the payoff diagram for a put option. Instead of purchasing a stock, the individual must sell it at a predetermined price. The dotted line represents the net payoff.

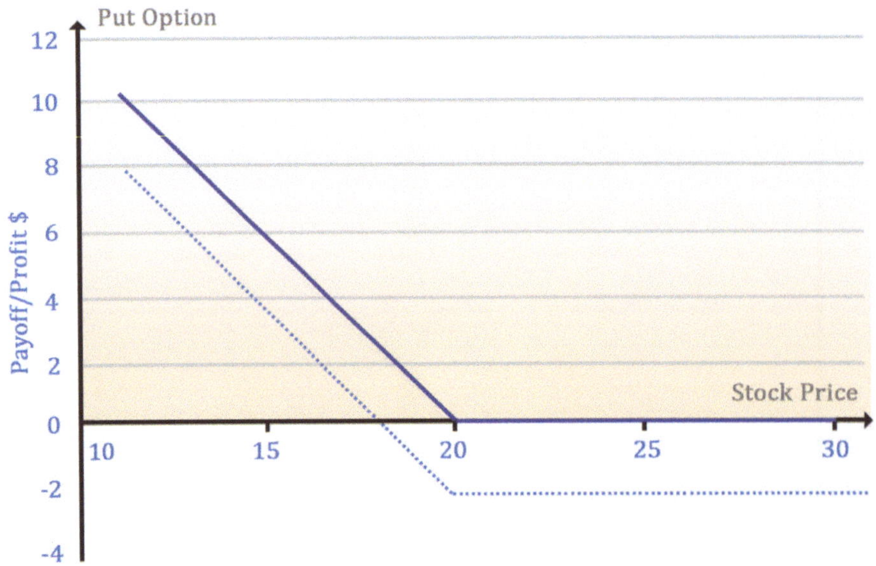

Figure 19: Put Option

The rate of return for an option differs from that of a stock. If the stock price fails to exceed the call option's strike price, the rate of return is negative 100%. The call option becomes worthless, and the individual loses everything.

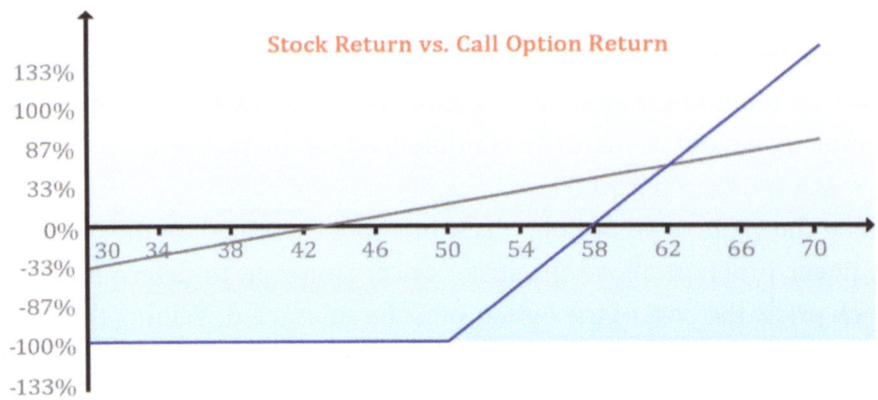

Figure 20: Stock Return versus Stock Option Return

However, as the stock price surpasses $50, the rate of return becomes positive, which is the rate of return for purchasing the stock. The graph illustrates that investing in an option provides leverage. This means that the investor receives an increase in the rate of return, as they receive the upside while protecting themselves from the downside. The worst-case scenario is losing everything, but this asymmetry means that there is a floor to how bad the investment can perform.

This nonlinearity is appealing to some investors who desire to take more risks. Biomedicine contains various embedded options, which implies that it is riskier than a typical investment that lacks these options. There are various option portfolios that one can construct to achieve specific risk-reward objectives. By analyzing the payoff diagrams of long or short positions in a given option, one can determine how to combine them to achieve the desired payoff.

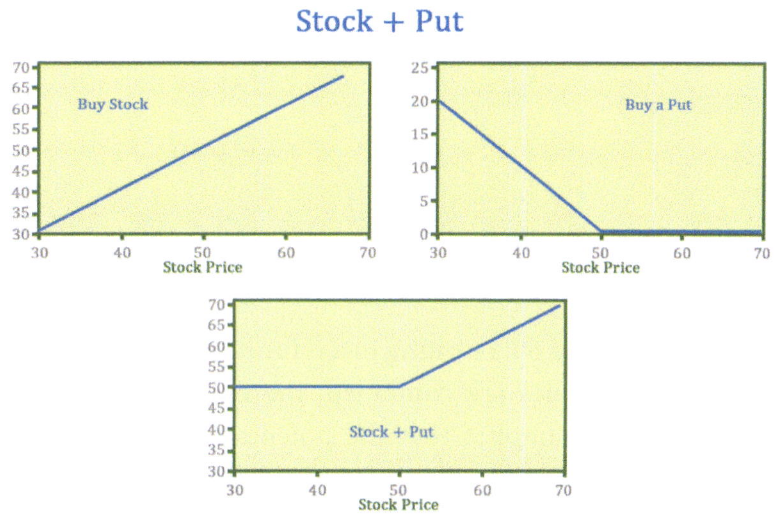

Figure 21: Stock Plus Put Option

A stock plus a put option on that stock at a predetermined strike price is an example of such a portfolio. It provides insurance that the investment's value will not decrease below a specific level. Options are akin to insurance, and the combination of purchasing a stock and a put option is a strategy that provides protection while allowing for an increase in the value of holdings if the stock price rises.

Here is an intriguing additional advantage. This entails purchasing a call option at a specific price and subsequently selling a call option at a higher price. By doing so, a graphical representation of the payoff is obtained.

It appears that both the downside and upside are limited. What is the rationale behind this? If there is a belief that the value will not surpass 60, it is unnecessary to pay for the potential gains beyond this threshold. If one firmly believes that the company's worth will never exceed 60, then why pay for that privilege?

By solely purchasing a call option, one is essentially agreeing to pay increasing amounts of money as the stock value rises, even if it surpasses 60. By dismissing the possibility of it exceeding 60, one is essentially paying for this potential. Why not eliminate this risk?

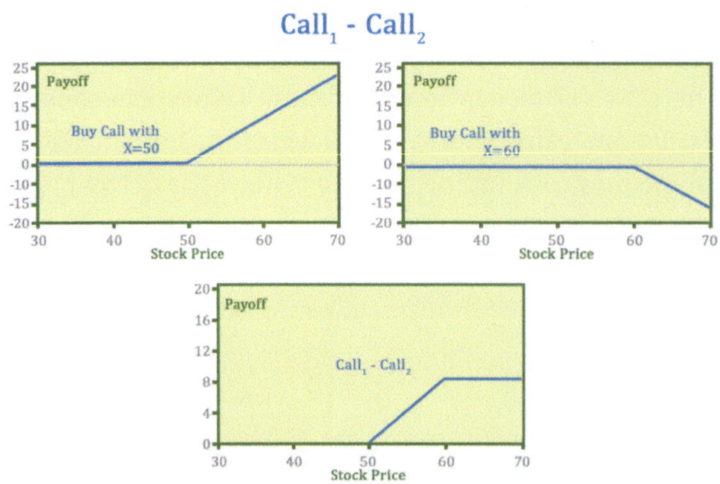

Figure 22: Selling a Call Option

The solution is to sell a call option at 60, resulting in a refund for any value exceeding this amount. Therefore, when these two strategies are combined, the cost is lower than solely purchasing the initial call option at 50. This implies that the potential gains are limited to 60, which is inconsequential if one does not believe it will exceed this value.

Consequently, one is content with this approach. However, if there is a belief that it can surpass 60, then purchasing the call option at 60 is a delightful opportunity. Thus, both parties are satisfied. This illustrates how options provide a means for individuals with differing perspectives to engage in mutually beneficial transactions.

Consider the potential if this approach could be applied to clinical trials. There are numerous disagreements regarding the success or failure of clinical trials, yet there are currently no convenient methods for speculating on their outcomes. Imagine the possibilities if options markets for clinical trials could be established. However, it is worth mentioning now since it is relevant to our discussion.

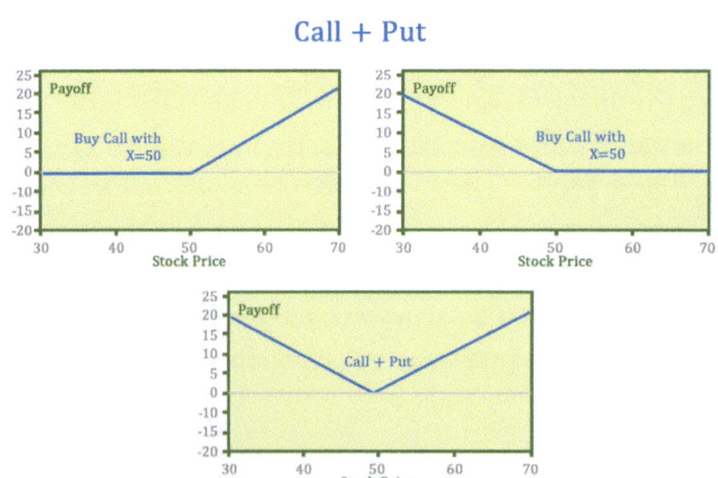

Figure 23: Call Option and Put Option at the Same Strike Price

Another example can be provided to illustrate the concept. Let us consider a scenario where a call option and a put option are purchased at the same strike price. This combination results in a V-shaped payoff diagram at that particular strike. The question arises as to who would be interested in buying such an option.

This type of bet involves uncertainty, as it is impossible to predict whether the underlying asset will increase or decrease in value. However, the expectation is that there will be significant volatility in the market. The risk of losing money arises when the prediction is incorrect, meaning that the stock remains stagnant without any significant movement.

This situation is unfavorable for the buyer. Conversely, if there is anticipation of major news, regardless of whether it is positive or negative, this strategy becomes appealing. The potential applications of engaging in these types of transactions are evident.

Moving on to pricing, it is worth noting that there are various financial instruments that resemble options without being classified as such. Despite this distinction, these instruments possess similar characteristics to options according to mathematical analysis.

Corporate Liabilities:

$$\begin{aligned}\text{Equity} &\equiv \text{Max}\,[0\,, V - B] \equiv E \\ &\equiv V - B + \text{Max}\,[0\,, B - V]\end{aligned}$$

$$\begin{aligned}\text{Debt} &\equiv \text{Min}\,[V\,, B] \equiv D \\ &\equiv B - \text{Max}\,[0\,, B - V]\end{aligned}$$

$$V = D + E$$

An example of this is corporate equity and corporate debt. Equity, in the context of a company, represents the residual value remaining after the bondholders have been paid off. If we consider a company's capital structure, where represents the amount of debt and represents the value of corporate assets, the equity is calculated as the difference between and .

However, if this difference is negative, the equity value is considered to be zero. Can anyone explain why negative equity is not possible? For instance, if the value of the assets is $100 million and the company has borrowed $110 million, why is not the equity value -$10 million? The answer lies in the concept of limited liability, which ensures that the value of equity cannot be negative. This principle of limited liability is indeed a remarkable aspect.

In the context of equity ownership, limited liability refers to the state in which the individual's responsibility for the company's debts is restricted. Consequently, the maximum loss one can incur is the entirety of their investment, which may not seem significant at first glance. However, it is crucial to recognize that there are more unfavorable scenarios.

For instance, in certain jurisdictions, bankruptcy proceedings may extend beyond the individual's assets and encompass their family members and children. Clearly, limited liability holds substantial value as it ensures that only the initial investment is at risk. Other personal assets remain safeguarded and cannot be seized to satisfy the specific liability.

This protection allows individuals to undertake risks without the fear of losing their homes or jeopardizing their future. Consequently, it grants them the freedom to pursue numerous entrepreneurial opportunities.

Moreover, limited liability bears resemblance to a call option when expressed as the maximum value between zero and the difference between the asset value () and the debt value (). Conceptually, equity ownership in a company can be likened to holding a call option on its assets, with the strike price being equivalent to the debt value.

If the outcome of this equation is negative, the equity owner is able to walk away unencumbered. In essence, equity holders assume greater risk compared to asset holders due to implicit leverage. By employing the put-call parity relationship, the leverage in equity can be observed, indicating that equity resembles a safeguarded, leveraged put option.

While this may seem complex, it is important to explore the accompanying notes and videos for a better understanding. The fundamental concept behind this structure is that corporate securities possess option-like attributes, which can be analyzed as if they were options.

Furthermore, there are other financial instruments that exhibit option-like characteristics, such as more intricate versions where the strike price is an average of multiple prices or the asset value is not a single figure but an average of various indexes. The subsequent discussion will focus on real assets that possess option-like features, specifically patents and tangible investment prospects. These assets, unlike financial securities, offer unique qualities that resemble options.

A patent does not serve as a form of financial security. However, it possesses a characteristic similar to that of an option, granting the holder the right, though not the obligation, to partake in a specific activity. Consequently, it holds value. Various other entities can also be perceived as options.

In fact, an individual once remarked that individuals with a particular expertise tend to perceive everything as an opportunity, likening them to someone with a powerful tool, such as a hammer, which can be utilized to accomplish a multitude of tasks. Therefore, we will examine in detail how this concept applies to specific types of applications.

Valuation of Options

In option pricing, the first names that come to mind are Black, Scholes, and Merton, who are rightfully credited for their contributions. However, the majority of option pricing applications have been executed using the binomial pricing formula, which is computationally simpler. The present work aims to introduce a simplified version of the formula that has practical applications.

Binomial Option-Pricing Model of Cox, Ross, and Rubinstein (1979):

- Consider one-period Call Option on stock XYZ
- Current stock price
- Strike price
- Option expires tomorrow,
- What is today's Option price ?

This version was developed by John Cox, Steve Ross, and Mark Rubinstein in 1979. The authors introduce a one-period world scenario, where an option is set to expire tomorrow, and the objective is to determine its value today. The authors assume that the stock price behaves like a Bernoulli trial, where a coin is flipped, and the probability of the coin flip is represented by p.

If the stock goes up, it increases by a factor u, and if it goes down, it decreases by a factor d, where u is greater than d. The authors aim to calculate the value of the option today (C_0) using this information.

Suppose $S_0 \rightarrow S_1$ Is A Bernoulli Trial:

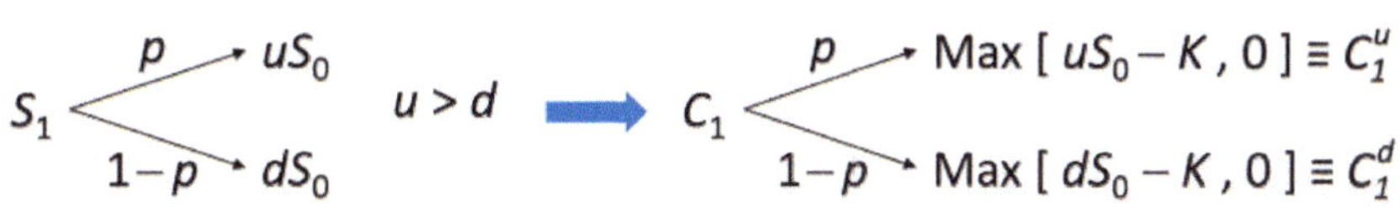

- For example, $u = 1.10$, $d = 0.90$ (S_0 moves up or down by 10%)

Now What Should $C_0 = f(\cdot)$ Depend On?

- Parameters: S_0, K, u, d, p, r

The statistical characteristics of the stock price are being discussed in this passage. Upon recording this information, it becomes evident that the value of the option for the following day is automatically determined. It is determined by the actual stock price.

If the stock price increases, the value of the option tomorrow will be the difference between US$0 and the maximum of that difference. This will be the value of the option in a favorable state of the market. Conversely, if the stock price decreases, the value of the option will be the value of the option in an unfavorable market state, denoted as . These two possibilities occur with probabilities p and 1-p, respectively. Are there any inquiries regarding this explanation?

If we assign u as 1.1 and d as 0.9, meaning that the stock price increases by 10% or decreases by 10%, respectively, then all of this information can be made more tangible. We can assign specific numerical values to the parameters. Therefore, the question I would like to pose is: what should be the present value of the option at date zero, considering this straightforward framework?

Now, let us consider all the various factors on which the value of the option could depend, including the stock price, exercise price, the factors for upward and downward movements, the probability, and the interest rate between today and tomorrow.

The formula for the current price of an option is independent of all these variables. Therefore, it is necessary to derive the formula for the current price of an option and understand why one of these variables is excluded. To illustrate this, let's consider the construction of a portfolio consisting solely of stocks and risk-free bonds.

- Hold Δ shares of stock, $\$B$ of bonds
- Total cost today: $V_0 \equiv S_0\Delta + B$
- Payoff V_1 tomorrow:

$$V_1 \begin{cases} \xrightarrow{p} V_1^u = uS_0\Delta + (1+r)B \\ \xrightarrow{1-p} V_1^d = dS_0\Delta + (1+r)B \end{cases}$$

Assume it is date zero. At date zero, the objective is to create a portfolio with a quantity of delta shares of stock and b dollars of risk-free bonds. This is achieved by purchasing delta shares of stock, which can be a positive or negative value, indicating the possibility of short selling. Similarly, b can also be positive or negative, representing the amount of borrowing. The cost of constructing this portfolio today, denoted as V_0, is given by the product of delta and the current stock price, plus B. This represents the total expenditure required for portfolio construction.

Now, let's consider the value of the portfolio tomorrow, assuming the stock behaves like a Bernoulli coin toss. The value of the portfolio tomorrow can be expressed as either $uS_0D + (1+r)B$ or $dS_0D + (1+r)B$, representing the two possible outcomes. This value is contingent upon the stock's behavior.

Now, it is important to note that the portfolio will be constructed in a unique manner. Specifically, D and B will be chosen to satisfy two specific conditions. In the favorable scenario, the value of the portfolio will precisely match the payoff of the call option in that scenario. Similarly, in the unfavorable scenario, D and B will be selected to ensure that the portfolio's value aligns exactly with the value of the call option in that scenario.

Now Choose Δ and B So That:

$$V_1 \begin{cases} \xrightarrow{p} uS_0\Delta + (1+r)B = C_1^u \\ \xrightarrow{1-p} dS_0\Delta + (1+r)B = C_1^d \end{cases} \Rightarrow \begin{aligned} \Delta^* &= (C_1^u - C_1^d)/(u-d)S_0 \\ B^* &= (uC_1^d - dC_1^u)/(u-d)(1+r) \end{aligned}$$

Then It Must Follow That (Why?):

$$C_0 = V_0 = S_0\,\Delta^* + B^* = \frac{1}{1+r}\left(\frac{(1+r)-d}{u-d}\,C_1^u + \frac{u-(1+r)}{u-d}\,C_1^d\right)$$

Can this always be achieved? Since we have two equations and two unknowns (D and B), and assuming they are not perfectly correlated or multiples of each other, basic algebraic principles guarantee the existence of a unique solution, denoted as D* and B*, that fulfills both equations. This solution is presented here.

The value of the portfolio should correspond to the price of the call option on the current day. In other words, the worth of the portfolio should be equivalent to the cost of constructing it at the beginning, as it effectively duplicates the future payout. Therefore, it is logical to question why anyone would pay a different price for the call option when the portfolio can replicate its payoff.

Suppose $C_0 > V_0$

- Today: sell C_0, buy V_0, receive $C_0 - V_0 > 0$
- Tomorrow: owe C_1, but V_1 is equal to C_1!

Suppose $C_0 < V_0$

- Today: Buy C_0, sell V_0, receive $V_0 - C_0 > 0$
- Tomorrow: owe V_1, but C_1 is equal to V_1!

$C_0 = V_0$, Otherwise Arbitrage Exists

$$C_0 = V_0 = S_0\,\Delta^* + B^* = \frac{1}{1+r}\left(\frac{(1+r)-d}{u-d}\, C_1^u + \frac{u-(1+r)}{u-d}\, C_1^d \right)$$

This claim can be substantiated through a contradiction. If the call option's value were higher than that of the portfolio, one could use the portfolio construction method and sell a call option to profit from the difference. Conversely, if the call option's value were lower, one could sell the portfolio and purchase the call option instead.

In both scenarios, the portfolio offsets the call option obligation, ensuring a certain profit. Consequently, constructing the portfolio correctly leads to an arbitrage opportunity, where the outcome is favorable regardless of market conditions. Hence, it is impossible for the option value to deviate from the value of the portfolio.

The Role of p

This formula possesses a compelling nature due to the absence of a particular parameter, namely the physical probability, p. The probability of the stock price increasing or decreasing is not accounted for in this formula. This may appear incorrect as one might expect p to be included. After all, if there is a high likelihood of being profitable, the option would be more valuable compared to a situation where profitability is uncertain. So why is p not included?

$$C_0 = V_0 = S_0\,\Delta^* + B^* = \frac{1}{1+r}\left(\frac{(1+r)-d}{u-d}\, C_1^u + \frac{u-(1+r)}{u-d}\, C_1^d \right)$$

$$= \frac{1}{1+r}\left(p^*\, C_1^u + (1-p^*)\, C_1^d \right)$$

$$C_0 = \frac{1}{1+r}\mathrm{E}^*[C_1] \qquad p^* \equiv \frac{(1+r)-d}{u-d}$$

The reason is that it is captured by the variables u and d. In essence, u and d serve as proxies for *p*. The question then arises: how do u and d fulfill this role? To explore this, let me demonstrate. Before delving into that, it is worth noting that there is a form of probability at play here. Allow me to illustrate this with the formula.

When different parameters are substituted into the formula, it takes on a certain appearance. In fact, it bears a resemblance to a discounted expected value. If I designate this particular quantity as a probability, denoted as p^*, it aligns precisely with the expected value of tomorrow's option payoff discounted at the risk-free rate. It is important to note that the expectation is not based on the physical probability, *p*, but rather on a peculiar pseudo probability that I have defined arbitrarily. Is *p* star truly a probability?

Firstly, it must always fall within the range of 0 and 1. Notably, for p^* to satisfy this criterion, u must be greater than 1 + r, and 1 + r must be greater than d. This condition holds economic significance as 1 + r represents the risk-free rate. If d exceeds the risk-free rate, no one would invest in bonds as the worst-case scenario for stocks would be superior to risk-free borrowing and lending.

Consequently, the risk-free asset would cease to exist. Similarly, if u were less than 1 + r, individuals would refrain from purchasing stocks since the best-case scenario would yield returns lower than risk-free borrowing. Hence, stocks would not be viable, and only bonds would remain.

Therefore, this can be considered a form of probability, albeit not the actual probability but rather a pseudo probability. Let's now see precisely how u and d incorporate *p* into this expression. As you already know, determining the appropriate cost of capital is essential for discounting a risky cash flow.

In the context of pricing stocks, the conventional approach involves adjusting the expected value of cash flows for risk by discounting it using the risk-adjusted cost of capital. The CAPM framework was developed to establish a method for making this risk adjustment. Therefore, if we adhere to this logic, the stock price will reflect this approach.

$$S_0 = \frac{1}{1+R} E[S_1] = \frac{1}{1+R} [\, p u S_0 + (1-p) d S_0 \,]$$

Today's stock price (S_0) is defined as the expected value of tomorrow's stock price discounted at the appropriate cost of capital. Hence, S_0 should be equal to 1 plus the risk-adjusted cost of capital (R). It is important to note that R represents the cost of capital adjusted for risk. The expected value of tomorrow's stock price, discounted at the risk-adjusted rate, is equivalent to today's stock price.

Calculating the expected value of tomorrow's stock price involves using physical probabilities, while discounting it with the cost of capital is encompassed within S_0. Therefore, if S_0 is used in an option pricing formula, the use of physical probabilities becomes unnecessary. Today's stock price encompasses all the necessary information for pricing the option relative to the stock.

Although this concept may seem complex, it is deeply rooted in economic theory. To simplify, consider the analogy of pricing a 6-ounce bottle of ketchup. If it is priced at $1.99, one can reasonably estimate that a 12-ounce bottle of ketchup would be priced around \$3.99 or \$4. The value of the 6-ounce bottle of ketchup is known, but the logic behind its price is determined by supply and demand.

Pricing an asset in isolation, without any reference to other factors, presents significant challenges. However, pricing an asset in relation to another asset is comparatively easier, as the groundwork of determining physical probabilities and risk-adjusted cost of capital has already been done. The valuation process has already established that a 6-ounce bottle of ketchup should be valued at \$2.

Consequently, based on this known value, it becomes possible to determine the worth of an 18-ounce or 24-ounce bottle, as well as other related products derived from the 6-ounce bottle. Some economists have attempted to diminish option pricing theory by labeling it as "ketchup economics," a term popularized by Larry Summers.

While it is true that option pricing is simpler, it encompasses much more than just the valuation of ketchup. In reality, option pricing provides a comprehensive framework for constructing various ketchup-like products, enabling accurate pricing and hedging strategies for businesses.

Thus, when examining the basic binomial option pricing formula, the inclusion of physical probabilities becomes unnecessary because the pricing is based on arbitrage. By correctly pricing today's stock prices, it becomes possible to accurately determine the prices of related items through the construction of an arbitrage equivalent.

For example, if a 12-ounce bottle of ketchup costs \$20, the logical course of action would be to purchase all the 6-ounce bottles, transfer their contents into 12-ounce bottles, and sell them at a price below \$20, thereby undercutting the market. This strategy can yield significant profits. The purpose of conveying this information is to highlight that the binomial pricing model can be expanded to multiple periods and ultimately evolve into the well-known Black-Scholes formula.

Multiperiod Extension

Multi-period option pricing is possible in the context of arbitrage arguments, as described earlier. The width of the binomial tree allows for the utilization of a seemingly simplistic coin toss to approximate intricate dynamics. By reducing the time steps and movements, the binomial tree can be employed at a microsecond level, transforming the approach into a continuous model that ultimately yields the Black-Scholes pricing formula.

Suppose We Let $n \to \infty$, $u, d \to 1 \pm 1/n$

- Continuous-time/continuous-state version (Black-Scholes/Merton):

$$\frac{1}{2}\sigma^2 S^2 \frac{\partial^2 C}{\partial S^2} + rS\frac{\partial C}{\partial S} + \frac{\partial C}{\partial t} = rC \ , \ C(S,T) = \text{Max}[S-K,0] \ , \ C(0,t) = 0$$

$$C(S,K,r,\tau,\sigma) = S\Phi(a_1) - K\exp(-r\tau)\Phi(a_2)$$

$$a_1 \equiv \frac{\log\left(\frac{S}{K\exp(-r\tau)}\right)}{\sigma\sqrt{\tau}} + \frac{\sigma\sqrt{\tau}}{2} \ , \ a_2 \equiv a_1 - \sigma\sqrt{\tau}$$

$$\Phi(\cdot) \equiv \text{Normal CDF}$$

Despite its simplicity and ease of implementation, the binomial method converges to Black-Scholes in the long run. Consequently, both methods can be utilized, but many individuals prefer the binomial formula due to its simplicity and ability to accommodate more complex strategies.

Real Options and Biomedical Examples

The fundamental concept is that an option, as the name suggests, grants the holder the right, but not the obligation, to engage in a certain action. For example, when establishing and operating a factory, an option allows the owner the right to manufacture goods but does not mandate the utilization of the factory. If necessary, the factory can be closed and left idle. Similarly, a patent operates on the same principle.

A patent is essentially a legal option that bestows upon the holder the right, but not the obligation, to utilize specific intellectual property. If others make use of this intellectual property, they are required to compensate the patent holder for the privilege. The patent dispute mentioned by both Harvey and Andrew pertains to the question of whether the right granted by a patent is the right to exclude or the right to use.

While it is true that a patent can be used to exclude others from using the intellectual property, it is also possible to license the patent to another party, granting them the right to use it. Fundamentally, a patent grants ownership of intellectual property and establishes a legal monopoly for a period of 20 years. After this period, anyone is free to use the intellectual property.

This reciprocal arrangement is a crucial aspect of patents. Interestingly, patents are entirely conceptual and not tangible. They are a construct of our legal system, established through a social agreement. Nevertheless, this legal construct holds genuine value in terms of monetary worth.

Although it lacks physical substance, it generates actual economic value. To illustrate this, we can examine the case of Editas, a company involved in a legal battle over a patent with the University of California. On February 15, 2017, the Broad Institute, which had licensed the patent to Editas, emerged as the victor, as announced.

By analyzing the stock price of Editas between February 14 and 21, during the unfolding of these events, we observe that the company's value increased by more than $300 million. Although ongoing litigation remains, the market perceived this particular decision as significant, attributing a value of over $300 million to it. Thus, if one were to inquire about the worth of a patent, one could consider it as an option that provides the flexibility to undertake actions that are currently beyond our capabilities.

This phenomenon is still in the preliminary stages of exploration. Nonetheless, it possesses significance. It should be noted that Andrew Hack, who will be introduced later in the discussion, served as the chief financial officer at Editas. Upon his announcement of departure, the company's stock price experienced a decline of approximately 14%. This indicates the extent of his value to the organization.

Consequently, a comprehensive body of literature emerged known as real option analysis, which extends the concept of a financial option to various non-financial contexts not originally encompassed by the Black-Scholes and Merton framework. The present discourse aims to apply real options using a different methodology, specifically decision tree analysis, which is more adaptable to non-financial settings.

The concept of optionality, i.e., the right but not the obligation to undertake an action, will still be utilized. However, we will employ a simpler framework that enables straightforward calculation of these variables, without relying on the complexity of the Black-Scholes formula.

Real Options

The concept of option analysis, when applied to real or corporate contexts, revolves around making specific investment decisions that are irreversible in nature. Unlike financial assets that can be sold to recoup the investment, irreversible investments, such as building a factory, do not offer the possibility of getting the money back.

This lack of reversibility is particularly relevant in the field of biomedical discovery, where decisions regarding clinical trials must be made at various stages. Each step in the clinical trial process represents an irreversible decision, akin to options that need to be evaluated and valued in a distinct manner.

For instance, consider the scenario of acquiring manufacturing facilities to develop a drug after obtaining FDA approval for marketing. In order to produce and distribute the drug, it becomes necessary to construct or acquire commercial manufacturing facilities that meet specific requirements. This was exemplified by Agilis Therapeutics, which successfully completed phase two of their gene therapy for AADC and received FDA permission to bypass phase three.

However, they still need to apply for a biologics license, which entails demonstrating to the FDA that they can produce therapeutics at a commercial grade level with proper quality control measures in place. The FDA conducts site visits to ensure compliance with safety criteria, and failure to meet these criteria will result in the denial of a biologics license.

Consequently, Agilis Therapeutics anticipates a year-long process from filing the license application to its approval, during which they must determine how to manufacture AAV at a scale that can meet patient demands. This necessitates a meticulous and rigorous process to ensure successful scalability.

In the scenario presented, a company has received approval for a new drug and must now consider acquiring manufacturing facilities. The drug is expected to generate profits of $200 million in the current year, but upcoming legislation may affect its price in the future. If the legislation is favorable, the price can be increased to $300 million, but if not, it will decrease to $100 million for the rest of the patent life.

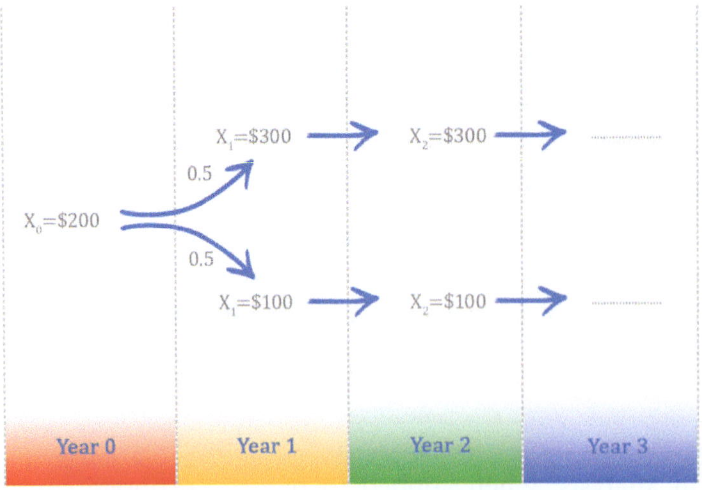

Figure 26: Legislation's affect on price

The decision to acquire the manufacturing plant is based on the risk-adjusted net present value, which must be calculated using the cost of capital. The cost of capital for the pharmaceutical industry is typically around 5%, but for this particular drug, which has already received FDA approval, a cost of 10% may be more appropriate.

$$\text{rNPV} \quad = \quad -1{,}500 \; + \; 200 \; + \sum_{k=1}^{\infty} \frac{0.5 \times 100 + 0.5 \times 300}{1.10^k} \quad = \quad 700$$

The risk-adjusted NPV is calculated based on the cost of acquiring the plant today, which is $1.5 billion, and the expected profits from the drug in the future. If the company acquires the plant today, the risk-adjusted NPV is $700 million. However, if they wait a year, the risk-adjusted NPV is $1.1 billion, which may be a more favorable option.

In the context of legislation potentially being unfavorable, the viability of a project worth $100 million is at stake. If the legislation goes against the project, it would become non-viable and essentially worthless. The reason for this is that in such a scenario, the project would only yield $100 million, which results in a loss considering the initial investment.

However, one does not necessarily have to operate the plant and can choose to keep it idle. The option of not using the plant holds value. The worth of this option is estimated to be $50 million, which is the difference between scenarios where the plant can be shut down in the event of unfavorable legislation. This demonstrates that the ability to shut down the plant adds value to it.

An alternative scenario is presented, where the project owner contracts with another party to produce the drug. In this case, the owner incurs expenses and loses the option of shutting down the plant. The concept of a real option is based on the idea that flexibility creates value.

The discussion continues with a question raised by Karthik regarding why the value of the option is only $50 million. The response explains that if the decision is made to proceed with the project, an additional $1,500 would be spent, resulting in a $50 million loss.

Furthermore, the plant would need to be repurposed for a more profitable use. Repurposing the plant would add further value to it, making the option even more valuable than the initial estimation. However, for the simplicity of the scenario, the assumption is made that the plant will be shut down and not repurposed.

The content then introduces a situation where the owner of the plant is willing to wait for a year before selling it. However, if the owner is not willing to wait, they offer the plant for sale at $1.5 billion, or it will be sold to another buyer on the open market. The question is posed regarding what actions can be taken in such a situation.

The individual could propose compensating him for the opportunity to wait for a year. What is the maximum amount you would be willing to offer? $50 million. Typically, it is not necessary to offer $50 million. If you were to pay him, let's say, $10 million just to remain idle and take a one-year break, it is likely that they would be receptive to it, unless, of course, there are other individuals competing for the property.

In that case, a more thorough analysis is required. However, this concept is the foundation of real options. It is the understanding that having the flexibility to decline certain projects and the various milestones that enable you to terminate them holds real value. It is important to consider and incorporate this into your calculations.

$$\text{rNPV} = \text{E}\left(\text{Max}\left[\frac{-1,650}{1.10} + \sum_{k=1}^{\infty} \frac{X_k}{1.10^k} , 0\right]\right)$$

One way to identify the presence of an option in your decision-making process is to express it mathematically, where a maximum function should be included in your formula. The maximum of two options allows you to select the more favorable one. This is similar to actual stock options.

The payoff of a typical call option is the maximum difference between the stock price at the maturity date and the strike price, with a minimum value of 0. So, the concept being discussed here is quite comparable. Therefore, when you calculate the net present value (NPV) and take its expectation, you will obtain the value of your project. The embedded option in this context is what introduces non-linearity and adds value.

You can expand this to include numerous periods using a binomial tree. The initial price of the asset is set at 200, but there is uncertainty regarding how high or low the price can fluctuate. Regardless of the binomial tree and probabilities you choose to employ, you can always assess the value of your assets based on this tree. The computation is straightforward and can be easily programmed into a spreadsheet. Although it may appear complex, it is not.

Decision Trees

In this scenario, a new drug has successfully completed phase I clinical trials, and the next step is to determine its effectiveness in a small population of patients through a phase II trial. The cost of this trial is estimated to be $5 million, with a 40% probability of success. If phase II is successful, the drug will require a larger phase III trial costing $50 million, with a high probability of success at 70%.

Upon successful completion of phase III, the drug will be approved by the FDA and worth a net value of $150 million. To calculate the NPV of future cash flows, a discount rate of 20% is assumed. A decision tree is drawn to analyze the various steps and probabilities involved in the drug's development, with year zero representing the successful completion of phase I. The expected NPV of the drug's worth is calculated at each milestone, taking into account the risk-adjusted cost of capital.

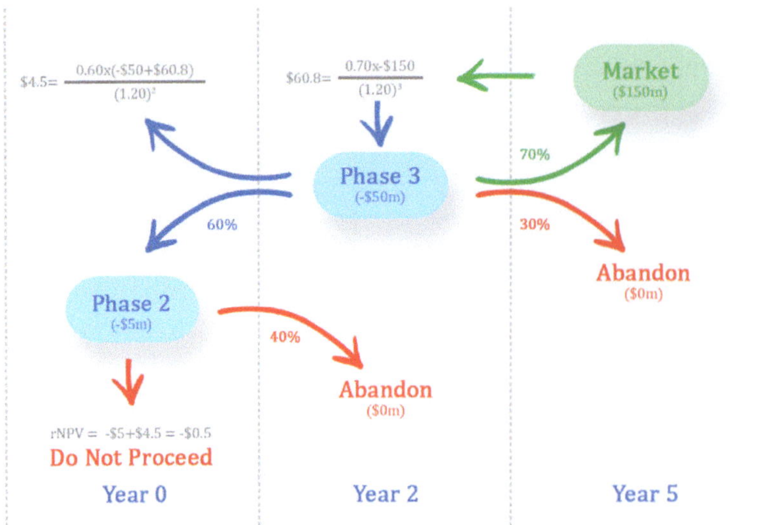

Figure 27: Decision Tree Stage 1

The cost of the phase III trial is estimated to be $50 million. The net value of the phase III trial, after deducting expenses, is approximately $11 million. Multiplying this value by 0.6 and dividing it by 1.2 squared, as of the desired date, yields a worth of $4.5 million. Considering the $5 million cost and the risk-adjusted net present value of $4.5 million, it is evident that this project would result in a loss of $0.5 million. Therefore, you are advised against pursuing this project, emphasizing that the ability to terminate it adds value.

The concept of option pricing theory is not utilized in this analysis, which is solely based on risk-adjusted expected value using a decision tree. In the case study of Merck, which will be examined during the week, the same approach will be followed, without the need for option pricing analysis. However, the underlying idea is that eliminating downside risks and the ability to terminate the project contribute to creating value.

The option pricing framework can be applied to clinical trials. These trials involve data collection and can have different outcomes in phase II, ranging from highly successful to mediocre. Depending on the outcome, the value of the phase III options varies. If phase II is highly successful, the phase III options have a 70% chance of being worth $350 million and a 30% chance of being worth nothing, which is an improvement compared to the previous scenario.

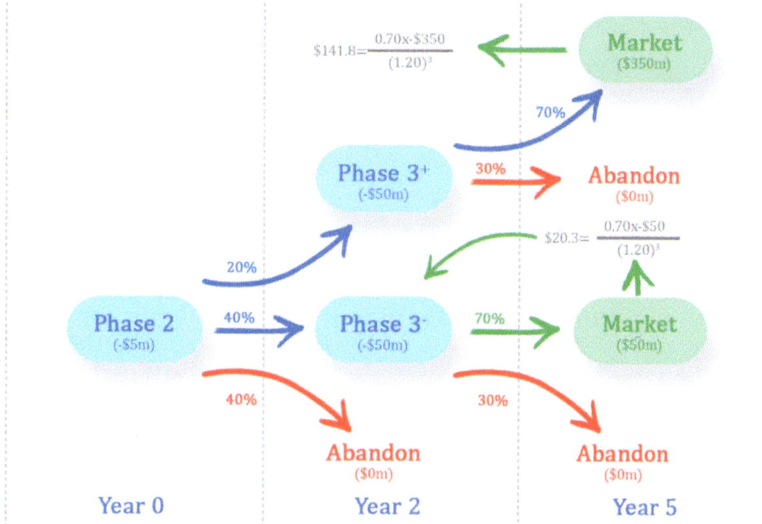

Figure 28: Decision Tree Stage 2

If phase II has yielded mediocre results, it can be asserted that phase III will likely have a value of $50 million compared to zero. This is because if the treatment effect is not significant in phase II, it is unlikely to improve in phase III. There is a possibility that the poor results in phase II were due to a small sample size, but it is more probable that the outcome of phase III will not be as successful.

Suppose we actually obtained this information instead of just one read-out of phase II, which indicated a value of $150 million or zero. In this case, there are two different possibilities for phase III. To analyze this, one must calculate the expected risk-adjusted cost of capital and the discounted value. By doing so, we can observe two distinct outcomes. When phase II is successful, phase III is valued at $91.8 million.

Figure 29: Decision Tree Stage 3

Conversely, when phase II is mediocre, the net present value (NPV) of phase III is negative \$29.7 million. Therefore, it is advisable not to proceed with the phase III trial if there is not strong evidence from phase II. This example demonstrates how real options can enhance value.

By eliminating the previous scenario and setting it to zero, the value can be recalculated. This involves determining the maximum value between the two possible outcomes for phase III, depending on the performance of phase II. Consequently, it can be seen that this approach can create real economic value. The project's worth increases to \$12.7 million compared to a negative \$0.5 million, solely due to the utilization of additional information.

Figure 30: Decision Tree Stage 4

It is important to note that decision trees can become complex, with multiple decision points, readouts, and interpretations. However, despite their intricate appearance, decision trees are actually quite systematic. The advantage of using decision trees is that it facilitates rational discussions between scientists and business professionals regarding whether or not to advance a particular trial.

Although the scientific merits may be compelling, if the NPV is negative \$29 million, it raises the question of who will finance the project. This does not imply that it should not be funded, as there are instances where large pharmaceutical companies have undertaken negative NPV projects for other reasons. Merck's decision to invest in and effectively cure river blindness using a drug with a negative NPV is a notable example. They did so for ethical reasons and because they had the necessary technology and knowledge.

Therefore, this framework enables a discussion on the costs for both the individual and the firm. The answer may not be favorable, but at least it provides a rational approach to the conversation. In conclusion, having access to additional information allows for informed decision-making and progression.

The value of optionality, as understood in this context, lies in its ability to eliminate a financial deficit of \$30 million. This is possible because one can obtain a more accurate prediction of the likely outcome. Another significant point to consider is that decision trees can be effortlessly established using Microsoft Excel. This aspect holds significance since it allows for the practical implementation of these concepts.

Furthermore, examining each column of the decision tree nodes can provide a potent method for engaging in business discussions and seamlessly integrating scientific insights into business considerations. Achieving this through alternative means can prove to be quite challenging. At present, there are no other tools that can match the ease with which all these diverse elements can be harmonized.

Monte Carlo Simulations

Monte Carlo, renowned for its appeal to high-stakes gamblers such as James Bond, derives its name from its reputation as a thrilling destination. The name itself embodies the essence of the city's allure. Although it may appear mundane, Monte Carlo possesses an extraordinary capability to simulate randomness across diverse domains.

Building on the foundation of decision trees, it becomes apparent that binary outcomes can be effortlessly modeled. However, the challenge arises when confronted with non-binary outcomes. For instance, when the precise effectiveness of a drug or the potential revenues are unknown, a broader representation as a random variable, such as a normal distribution, becomes necessary.

This concept, aims to calculate the influence of randomness on decision-making, particularly in relation to adjusted net present value. The objective is to develop an approach that is straightforward and easily comprehensible, ensuring effective communication with shareholders, investors, and other stakeholders. Hence, the efficacy of the tool lies not only in its sophistication but also in its interpretability. Monte Carlo simulation proves to be an exceptional method for encompassing these multifaceted aspects.

To introduce the fundamental concept of Monte Carlo simulation, it is essential to understand its simplicity. The initial step involves identifying the variables subject to randomness, such as revenues, drug efficacy, or patient enrollment. These uncertain variables, which significantly impact adjusted net present value, are then modeled as random variables.

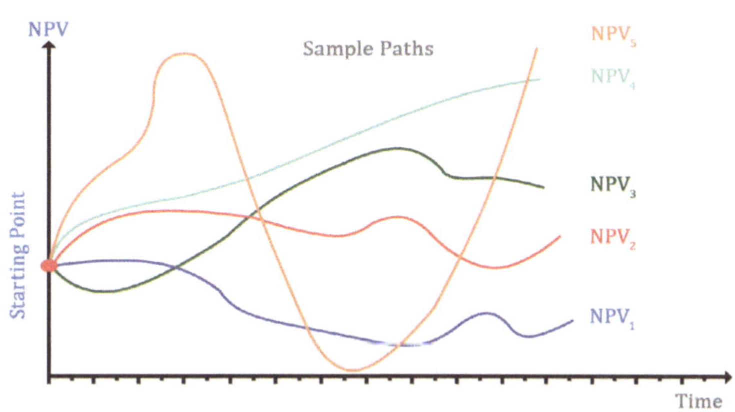

Figure 31: Random Variables

Various distributions, including normal, lognormal, gamma, and beta, are available for selection, assuming prior exposure to these concepts. Once a specific random variable is proposed, a decision rule is formulated to guide decision-making based on the characteristics of that variable.

- Suppose we model a project as a sequence of cash flows $\{X_t\}$, but where cash flows depend on decisions $\{Y_t\}$ and random variables $\{Z_t\}$:

$$X_0(Y_0, Z_0)\ ,\ X_1(Y_1, Z_1)\ ,\ \cdots\ ,\ X_T(Y_T, Z_T)$$

$$\mathrm{rNPV} \;=\; X_0(Y_0, Z_0) + \frac{X_1(Y_1, Z_1)}{(1+R)} + \cdots + \frac{X_T(Y_T, Z_T)}{(1+R)^T}$$

- Each sample path and sequence of decisions leads to a realized rNPV; now do this many times to get multiple sample paths

In the context of drug efficacy evaluation, if the initial experimentation with animal models reveals a drug efficacy greater than half a standard deviation, the next step is to proceed with IND filing. Conversely, if the efficacy falls short of this threshold, the drug will be withdrawn. This decision-making process is an example of a rule that is determined by random variables.

$$\begin{array}{lcccc}
\text{Sample path 1:} & X_0^1\ ,\ X_1^1\ ,\ldots,\ X_T^1 & \Rightarrow & \mathrm{rNPV}^1 \\
\text{Sample path 2:} & X_0^2\ ,\ X_1^2\ ,\ldots,\ X_T^2 & \Rightarrow & \mathrm{rNPV}^2 \\
\vdots & \vdots \quad \vdots \qquad\quad \vdots & \vdots & \vdots \\
\text{Sample path } m\text{:} & X_0^m\ ,\ X_1^m\ ,\ldots,\ X_T^m & \Rightarrow & \mathrm{rNPV}^m
\end{array}$$

Subsequently, a Monte Carlo simulation is employed to explore various outcomes. In this simulation, random variables are drawn, and their corresponding outcomes are applied to a specific set of decisions, resulting in a financial outcome. The outcome could be a significant loss of $20 million or a gain of $50 million, depending on the specific decision made. The Monte Carlo simulation is repeated numerous times to generate a distribution of outcomes. This distribution can then be used for various analyses.

For instance, the mean of the distribution represents the expected adjusted net present value, which can be used as a basis for decision-making. If the mean is positive, the project is deemed favorable, whereas a negative mean suggests that the project should be abandoned. Additionally, Monte Carlo simulation allows for the calculation of other statistical measures, such as the probability of incurring financial losses. This is particularly valuable as it provides insights beyond the expected value calculation. Furthermore, Monte Carlo simulation enables the assessment of the range between the best- and worst-case scenarios, providing a measure of risk.

By considering the potential upside and downside, investors can evaluate the level of risk associated with a particular investment strategy. Monte Carlo simulation offers these analytical benefits, which are not easily achievable through decision trees. Overall, Monte Carlo simulation provides a comprehensive framework for evaluating and analyzing investment strategies, offering insights into expected values, probabilities, and risk assessments.

The basic outline of the concept is as follows. To provide a formal explanation of the process, notation will be used, followed by an example. A visual representation will be presented to illustrate the procedure. The x-axis represents time, while the vertical axis represents the NPV of a specific investment. The starting point, referred to as time zero, signifies the initial condition of the investment.

A simulation of the identified randomness will be conducted, incorporating a decision rule based on the observed randomness. This will yield the NPV of the first path. By running the randomness through the process, the NPV at the end will be obtained. Any queries regarding this process can be addressed.

Moving on, a second draw of the random distribution will be conducted. The simulation will determine the progression from Phase 1 to Phase 2, Phase 2 to Phase 3, and the presence of toxicity effects. These uncertain variables will be modeled using random variables. The resulting NPVs for phases three, four, and five will be obtained.

Some paths may exhibit extreme outcomes, which is the purpose of conducting these simulations. The simulations will generate a collection of NPVs, which is the output of a Monte Carlo simulation. These numbers are of significant importance as they can be averaged to obtain the average NPV and analyzed to determine the extent of variability.

For instance, the spread of the NPVs appears substantial, although the scale has not been provided. If the lines representing the NPVs were closely aligned, it would instill more confidence in making decisions based on this process. Currently, the simulations suggest that any outcome is possible, which in itself provides valuable information. It indicates that the randomness incorporated in modeling each decision point has a substantial impact on the final NPV.

Now, let's express this concept in mathematical notation and proceed with its practical application. Suppose a project is modeled as a sequence of cash flows, with the cash flows being random and dependent on both your decisions and underlying state variables. These cash flows are expressed as a function of two variables, Y and Z.

The variable y represents the decision variables that you control, while Z represents the states of nature, such as drug approval, endpoints achievement, and toxicity effects. By representing the project in this manner, it can be demonstrated that the risk-adjusted NPV is determined by the sequence of decisions and the sequence of state variables (Y's and Z's).

In order to simulate a single sample path, it is imperative to generate a variety of random variables, such as Z_0, Z_1, Z_2, up to time t. Subsequently, one must employ a rule of thumb or decision rule to determine the appropriate choice based on the value of each random variable. For instance, if Z_0 is obtained, Y_0 is selected, and so on.

The underlying concept behind a Monte Carlo simulation entails running multiple iterations of these random variables, assessing the outcomes of each decision based on the realizations of the random variables, and calculating the risk-adjusted NPV. This process is then repeated multiple times.

Each sample path refers to a specific sequence of random variables for a given draw, representing an entire historical path. For instance, one realization could involve succeeding in Phase 1, succeeding in Phase 2, but failing in Phase 3. Another realization might involve succeeding in Phase 1, but being stopped in Phase 2.

Yet another realization could involve succeeding in all three phases and gaining approval, but experiencing poor revenue performance due to a market downturn, and so on. Each of these sequences provides a distinct replay of historical events. Once the consequences of these histories are calculated, a collection of m realized risk-adjusted net present values (RNPVs) is obtained. These m RNPVs form a distribution that can be analyzed as desired.

Based on these realizations, various statistics can be computed, including the expected value (i.e., the average), standard deviation, minimum, maximum, and other relevant measures. These statistics serve as the primary outcomes of the simulations and can guide decision-making processes once the expected value and standard deviation are determined.

Monte Carlo Simulation Example

Let us now proceed with an example. The concepts we have covered thus far may appear straightforward, but there are certain nuances that we will address shortly. Consequently, the anticipated value of the risk-adjusted net present value (NPV) can be approximated by the sample mean of a multitude of realizations, whether it be 100, 1,000, or even a million. By taking an average of these realizations, one obtains an estimation of the theoretical population expectation of this random variable.

Typically, this theoretical estimation is not easily calculable. However, through the process of simulation, one simply needs to conduct 100 or 1,000 trials, calculate their mean, and thereby acquire a reasonably accurate estimate. The accuracy of this estimate will be further explored shortly.

The key point is that when the average is computed across these simulations, it serves as an estimator for the theoretical expected NPV. In the same manner that the average of any variable is intended to estimate the population expectation, this average across simulations fulfills a similar role.

Now, we can proceed to estimate the variance of the risk-adjusted net present value by calculating the sample variance and so forth. Remarkably, these estimators are generally quite reliable. This reliability can be attributed to the assumptions underlying typical statistical estimation, namely the presence of independent and identically distributed random variables. As we are simulating independent and identically distributed random variables, these assumptions hold true.

In the aforementioned scenario, the process of multiplying the standard deviation by approximately 2 is discussed. The specific value mentioned is 1.96 or 2. This calculation is performed in relation to a standard deviation of 50 million, resulting in a product of 100 million. It is established that there is a 95% probability that the outcome will fall between 100 and 300. This approach bears similarity to the previous analysis involving the Bernoulli trial, with the advantage of accommodating various intermediate values.

The decision at hand pertains to the purchase of a facility, with the cost amounting to 1.5 billion presently or 1.65 billion in the following year. The question arises whether it is preferable to make the purchase immediately or to delay it.

The assumption is made that a discount rate of 10% is applicable. However, the present inquiry differs from the previous one, as it focuses on the likelihood of incurring financial losses in this transaction. Unlike the previous scenario, which only presented two possibilities, this situation entails a more nuanced query.

Furthermore, if the decision is made to postpone the purchase for a year, what is the probability of surpassing the NPV of an immediate acquisition? Additionally, if the standard deviation of profits is 150 million instead of 50 million, signifying greater uncertainty regarding potential profits, what implications does this hold for the decision-making process?

To address these questions, a simulation exercise is proposed. The objective is to generate 100 iterations of net profit realizations, represented by the random variable "x." This variable possesses a mean of 200 million and a standard deviation of 50 million. Each realization is then substituted into an equation to calculate the risk-adjusted NPV. After conducting this process for 100 iterations, an analysis of the resulting data is performed to inform the decision-making process.

The practical implementation of this exercise can be accomplished using Excel. Each of the 100 cells represents a random variable ranging between 0 and 1, obtained by taking the inverse of the normal distribution function. This conversion can be easily executed using an Excel function. Thus, the simulation exercise can be conducted within the Excel platform.

Upon completing the simulation with the 100 random variables, the distribution of risk-adjusted NPV is presented in the form of a histogram. The mean value is 200 million, while the standard deviation remains at 50 million. The simulation consists of 100 trials, resulting in the observed distribution.

Calculations reveal that the expected risk-adjusted NPV amounts to 487 million. Based on this figure, it can be concluded that pursuing the deal is advisable. Furthermore, the probability of incurring a loss can be determined by examining the realized values of risk-adjusted NPV that fall below zero. In this case, the probability of loss is precisely 21%.

With this simulation, the likelihood of losing money can be determined based on the given assumptions, which amounts to 21%. Additionally, there is a one-third probability of exceeding a value of 700, which is considered favorable. This simulation provides valuable information, although it is important to note that it is not a magical solution.

The information provided is dependent on the inputted assumptions, particularly the normal distribution with a specific standard deviation. If these assumptions are incorrect, the resulting inferences will also be inaccurate. It is crucial to remember the principle of "garbage in, garbage out." This highlights one of the challenges of simulation, as it requires accurate knowledge of the data-generating process.

Now, let us address the final part of the question regarding the impact of changing the standard deviation from 50 to 150. In this case, the net profit for the following year, represented by the x variable, exhibits a different pattern.

The simulation was conducted with a standard deviation of 150, resulting in a wider and more scattered distribution, as shown in the histogram. This is in contrast to the previous simulation, which had a narrower and more organized distribution. Both simulations are presented on the same scale for a fair comparison.

Now, let us consider a scenario where the real world becomes more uncertain, possibly due to factors such as changes in the legal infrastructure or political dysfunction. In such a case, the expected risk-adjusted net present value (NPV) experiences an almost twofold increase.

There is an increased number of outliers present. However, there are outliers both on the left and right side. The relationship between risk and return is positively correlated, especially in equilibrium models. The absence of equilibrium in this scenario is considered unusual. In the context of a CAPM model, it would be accurate to state that higher risk leads to higher returns.

Nonetheless, the absence of any CAPM model is recognized. The assumption of a CAPM model was not made. The reader is commended for being on the right track, but the explanation is simpler and relates to the calculation method being utilized. This is directly linked to the NPV. Indeed, the risk is adjusted, but this adjustment is specifically related to the option to delay the purchase of the plant until the following year. This optionality factor contributes to the boost in value.

The concept of real options is evident here, highlighting the benefits of being able to wait. Furthermore, it suggests that uncertainty is welcomed and randomness is desired in such situations. Ultimately, the risk-adjusted net present value is nearly doubled, but the variance has more than doubled.

The probability of loss is 45%, indicating a significant chance of losing money. Despite the higher mean, the risk of loss is almost half, which may deter investors. However, the likelihood of making more than 700 million on this deal is now 40%, which is greater than the previous 32%.

The decision to invest in such a project ultimately depends on individual risk preferences. In a hypothetical scenario where two competing projects are presented, one with a higher probability of loss and the other with a higher potential gain, individuals in the drug development industry tend to be more risk-averse due to the high stakes involved. This differs from other industries where financial gains are the primary focus. It is essential to consider the psychology of drug development when making investment decisions and consulting with stakeholders in the industry.

In the realm of Monte Carlo simulation, determining the appropriate number of simulations is a crucial factor. The optimal number of simulations is dependent on various factors, and it is not a straightforward answer. The central limit theorem is a fundamental concept in this context.

The estimation of the expected risk-adjusted NPV is a critical factor in Monte Carlo simulation. The Bernoulli trial, which is a single random variable that can be either 100 or 300, is important to understand. The steps involved in calculating the expectation considers the two possibilities and weighting them by their respective probabilities.

In order to assess this expectation explicitly, it can be done if a Bernoulli trial is available. However, if it is stated that the distribution of this variable, denoted as x, follows a complex gamma distribution, the calculation process becomes challenging and not straightforward.

This becomes even more complicated when there are multiple decision points involved. For instance, after phase 1, there is a realization of the compound's value, followed by another realization after phase 2, and so on, resulting in a sequence of three random variables.

Decisions are then made based on these realizations. Consequently, it becomes increasingly difficult to analytically calculate this expectation within a short period of time. A solution cannot be obtained.

Nevertheless, Monte Carlo simulation provides a means to estimate it by conducting numerous experiments and averaging the results. It is known that with a sufficiently large sample, the average value will converge to the theoretical value, as stated in the central limit theorem and the law of large numbers.

The law of large numbers indicates that the estimation of this expectation converges to the theoretical expectation when the number of simulations is large enough. However, the central limit theorem goes even further by stating that if the difference between the estimate and the true value is multiplied by the square root of m, where m represents a sufficiently large value, this quantity will converge to a normally distributed random variable with a mean of 0 and a variance of sigma squared rNPV.

This concept is known as the central limit. The lowercase "a" in this context represents asymptotic, which signifies the limit as m approaches infinity. This is significant because it allows us to determine the standard error of this estimator, which is a random variable with its own standard deviation, referred to as the standard error. The standard deviation is calculated by dividing sigma rNPV by the square root of m.

To clarify, there exists an estimator, denoted as q, for the rNPV (risk-adjusted net present value). q is the estimator of the true expected rNPV sought. Determining the value of q is needed to assess the accuracy of the estimator. It is therefore essential to determine the value of q. Using the central limit theorem, that the standard deviation of q can be estimated by dividing the standard deviation of individual simulations by the square root of m.

Although the exact value of this standard deviation is unknown, it can be estimated. By utilizing the available resources and distribution of the NPV, the variance of a set of 100 realizations can be calculated, which serves as an estimate.

Suppose the individual is preparing to meet with the CEO and present a project with an anticipated value of 400 million, but a standard deviation of 100 million. This implies a range of 400 million plus or minus 200 million. However, this level of uncertainty is deemed too significant for discussion with the CEO, indicating that an insufficient number of simulations were conducted. To address this, the number of simulations should be increased.

If the standard deviation is reduced from 100 to 10, the factor of reduction would be 10. To determine the additional number of simulations required to achieve this reduction, the current number of simulations (m) needs to be increased by a factor of 100, resulting in 10,000 simulations. With this increased sample size, the projected value of the project would be 400 million plus or minus 20 million, with a 95% probability.

For even greater precision, conducting a million simulations would yield a range of 400 million plus or minus $0.25, although such precision may not be realistic in practical scenarios. To determine the desired level of accuracy, it is necessary to estimate the standard deviation and calculate the sample size accordingly. The desired degree of precision determines the number of simulations to be performed.

These calculations are approximations based on the central limit theorem, but since the simulations are independently and identically distributed, they provide excellent approximations. Therefore, a relatively small number of simulations is sufficient to apply these results effectively. The process of conducting simulations, interpreting their outcomes, and determining the appropriate number of simulations can now be applied in various contexts.

Simulations prove to be an invaluable and efficient tool for obtaining prompt answers compared to the complexities and calculations involved in relying solely on decision trees, especially in cases involving intricate negotiations, license deals, milestone payments, and contract clauses. However, simulations require a clear understanding of the underlying uncertainty's nature, such as whether it follows a normal, log-normal, or beta distribution, as these assumptions impact result interpretation.

7. VENTURE CAPITAL & NEW BUSINESS MODELS

This chapter provides an in-depth analysis of the pivotal role venture capitalists play in offering financial support and guidance to fledgling companies. Within the healthcare sector, Venture Capitals (VCs) frequently finance significant milestones, such as clinical trials and product launches, aimed at enhancing the value of companies for potential acquirers and the public markets.

The chapter commences with an industry overview, followed by an examination of the essential elements of a term sheet, a document outlining the financial arrangement between the VC and the entrepreneur. It then delves into the methods by which VCs construct and evaluate valuation models for early-stage biopharma assets and companies. Lastly, it introduces the concept of venture philanthropy, a novel business model that applies venture capital strategies to philanthropic objectives.

Defining Venture Capital

One of the key aspects to understand in order to start a biotech company are the different perspectives involved. These perspectives include that of scientists and clinicians, entrepreneurs, investors, regulators, and then patients.

In the context of oncology, the acquisition of an approved medication represents a substantial asset valued at approximately \$12.3 billion. This signifies that following a decade-long period dedicated to clinical trials, successful approval of the drug leads to an asset yielding an annual profit of around \$2 billion from the 11th to the 20th year. When considering the present value, this amounts to approximately \$12.3 billion. This substantial potential return on investment serves as a key motivator for potential investors.

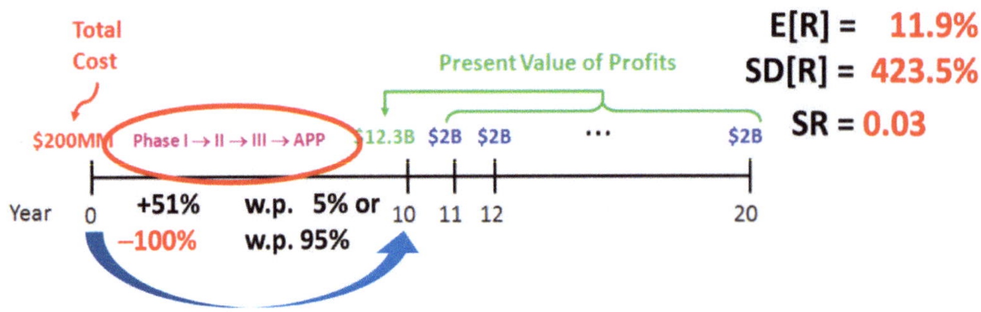

However, a critical analysis of the financial metrics reveals that an expected return of 12% and a standard deviation of 423.5% result in a Sharpe ratio close to zero. Consequently, the investment appears unattractive without additional mitigating circumstances, primarily due to the extended time horizon associated with the clinical development process. The protracted nature of progressing from phase one to phase three and ultimately obtaining a new drug application presents a significant challenge, as few investors possess the financial endurance to sustain prolonged periods of capital exposure without immediate returns.

Consequently, the industry has adopted a strategy of subdividing the 10-year clinical development process into smaller, more manageable investments, a practice in which venture capital plays a pivotal role. Venture capital represents a segment of the financial landscape dedicated to facilitating the transition of innovative concepts from the laboratory to clinical application and eventual regulatory approval. This specialized form of investment is crucial in supporting the developmental trajectory of early-stage ventures, providing the substantial financial infusion necessary to bridge the gap towards achieving positive cash flows.

In the context of startup enterprises, the initial phase involves the transformation of conceptual ideas into tangible products or services, necessitating significant financial outlays. During this period, the cash flow trajectory typically reflects a negative trend, signifying substantial expenditure before reaching a point of profitability. This developmental timeline varies across industries, with technology companies often experiencing a relatively short period before achieving positive cash flows, while biotechnology firms encounter protracted timelines before reaching significant milestones.

Notably, venture capitalists are inclined to engage during the early stages of a venture, providing substantial financial support to propel the enterprise towards achieving positive cash flows. This early-stage investment serves as a crucial bridge, facilitating the transition towards profitability and potentially culminating in an initial public offering (IPO), allowing investors to divest their shares and realize their investments. The overarching objective of venture capital is to navigate startups from their inception to a stage where they can attract public investment and enable investors to exit their positions.

The concept underlying a venture capital firm is to identify and support ideas that have already demonstrated significant business potential, with the goal of developing them into sustainable, profitable enterprises in the near future. A venture capital firm consists of distinct components, including general partners who oversee project selection, work with entrepreneurs, and manage the fund's portfolio of assets.

Limited partners, on the other hand, provide financial backing to the venture, often relying on the expertise of the general partners in the field. The facetious definition of a venture capital fund humorously emphasizes its finite lifespan, typically lasting 6 to 10 years, during which the general partners contribute their expertise, and the limited partners provide financial resources.

The roles of the two groups differ significantly, with general partners assuming primary responsibility for the fund's management, while limited partners have minimal involvement beyond providing capital. Additionally, limited partners are legally shielded from liability related to the fund's activities, unlike general partners who can be held accountable for improper investments or financial misconduct.

Once established, a venture capital fund invests in various companies, with the performance of these investments directly impacting the returns experienced by the limited partners. The fund's lifecycle begins with the general partners raising capital from limited partners, creating a pool of funds to be gradually invested over several years.

The actual transfer of funds occurs as needed, with general partners calling upon the limited partners to provide the necessary capital for specific investments. As the invested companies progress, the limited partners eventually receive returns on their investments. However, venture capital entails high risk due to the asymmetry of information and control, leading to the potential for substantial losses.

Investors in venture capital funds must be prepared for the possibility of losing their entire investment, as these ventures can be highly unpredictable. Nonetheless, the strategy behind a venture capital fund involves diversifying investments across multiple opportunities, aiming to offset potential failures with a few successful ventures that yield significant returns.

To mitigate the possibility of failure, venture capitalists protect asymmetries of objectives, information, control, and so. These protections are highlighted in the term sheet, which outlines the terms under which investors invest and the terms under which the investment will be used to generate returns for the venture capital fund.

The term sheet serves as a comprehensive summary of various crucial elements. While a detailed analysis of the term sheet is beyond the scope of this discussion, it is important to highlight key aspects. Interested individuals can access term sheets online through the National Venture Capital Association's website, which hosts a repository of legal documents curated by seasoned investors, legal experts, and entrepreneurs. These documents are designed to facilitate the negotiation process with venture capitalists (VCs).

Table 9 - Term sheet

	Term Sheet Component	Description	Objective
1.	Staged commitment of capital	To the entrepreneur, VC provides capital in several rounds, such as seed stage and expansion	To control firm development and manage investment risk
2.	Convertible securities	Debt-like security that can be converted into common stock, e.g., convertible preferred	To allocate control to the party with more expertise, provide incentives for the entrepreneur and distribute cash flow rights.
3.	Covenants	Contract provisions to establish governance of firm	To control the rights and incentives for the entrepreneur
4.	Registration rights	To define exit options for the VC and entrepreneur	To obtain liquidity for the VC and entrepreneur

One can find a comprehensive template of a term sheet, covering a wide array of potential scenarios. Prior to embarking on a biotech startup, it is strongly recommended that individuals thoroughly review and comprehend this document. The complexity of financial jargon may initially seem daunting to some, but it is imperative to grasp these concepts, as signing a document without full understanding can lead to unintended consequences.

Furthermore, when engaging with VCs, individuals should exercise caution, particularly regarding non-disclosure agreements. VCs may have invested in competitors, and discussions with them should be approached prudently, especially when safeguarding proprietary information. While VCs typically refrain from signing non-disclosure agreements, the biotech industry is an exception. A template for a non-disclosure agreement can also be found on the aforementioned website, although it is important to note that the efficacy of such agreements ultimately depends on the integrity of the signatories involved.

Staged Commitment and Dilution

The concept of stage commitment and dilution can be succinctly illustrated using a basic example centered around a balance sheet, which serves to encapsulate the overall value of a given enterprise at a specific moment in time. A hypothetical scenario will be utilized to convey the essence of stage commitment and is as follows:

A significant discovery In the laboratory with potential commercial applications for a specific ailment has been made. At the outset of the research, the balance sheet reflects the value of the idea, which remains undocumented and solely in the mind of the researcher(s).

This initial idea represents the sole equity held by them, effectively making them the sole proprietor of the company, comprising the researcher(s) and the idea. To progress from this conceptual stage to practical implementation, the researcher(s) must undertake various actions. For instance, filing a disclosure with a technology licensing office and applying for a patent, a process that demands financial resources, time, expertise, and effort. If they lack these resources, they may need to collaborate with others that can provide the necessary support.

Consequently, the researcher establishes a collaborative relationship, necessitating the development of a somewhat personal connection. This is significant because it complicates the process of initiating a business. Once an idea and a patent are secured, the company's valuation changes from zero to a new value, denoted as , indicating the creation of value.

This added value is essential for attracting interest from potential investors and venture capitalists, who recognize the legal protection and exclusivity associated with the idea. Subsequently, the next phase involves securing substantial funding to establish a company based on the idea and patent, a process typically facilitated by venture capitalists.

These investors engage with the technology licensing office to select suitable projects and negotiate terms for establishing a company, licensing relevant technology from the university, recruiting necessary personnel, and providing expertise while serving on the board of directors. As a result, the original entrepreneur, such as the researcher(s), who initially possessed only an idea, now holds a smaller stake in a significantly larger enterprise.

Table 10 – Stage Commitment Balance Sheet

Date 0		Date 1		Date 2	
Assets	**Liabilities**	**Assets**	**Liabilities**	**Assets**	**Liabilities**
				Idea, Patent	Susan, MIT
Idea	Susan	Idea, Patent	Susan, MIT	Cash, Plant	Investors
				Equipment	Employees
V_0	V_0	V_1	V_1	V_2	V_2

$$V_0 \overset{?}{<} V_1 \overset{?}{<} V_2$$

The primary challenge throughout this progression is effectively managing this process, which is the focal point of further discussion. The aim is to ensure that each stage, denoted as to to , results in the creation of value, with each subsequent value being greater than the previous one. The focus then shifts to the perspective of the entrepreneur, who, despite the concept of dilution from the venture capitalist's viewpoint, may find staged capital and dilution beneficial in certain contexts.

For instance, if one requires $8 million to advance a company, and the initial company valuation stands at $10 million, raising the necessary capital would result in the entrepreneur owning a reduced percentage of the company. Dilution holds significance not only from an economic standpoint but also in terms of control over the enterprise.

If the entrepreneur's ownership stake falls below 51%, they effectively become employees of another company. Many entrepreneurs find this transition challenging as they prefer to maintain control. However, in the biotech industry, it is exceedingly difficult for entrepreneurs to retain control due to the substantial financial resources required. Securing a large investment, such as $8 million, can result in a loss of control over the company.

Nevertheless, an alternative approach is available. Instead of immediately accepting the full amount, entrepreneurs could consider a staged capital commitment, initially taking a smaller sum, such as $2 million, and then seeking the remaining $6 million at a later stage when the company's valuation is higher. This approach allows entrepreneurs to retain a greater share of equity and potentially maintain some degree of control.

Despite its advantages, this strategy does come with a significant downside, particularly in the biotech industry. There is a risk that the required $6 million may not be accessible when needed, potentially leading to a disruption in operations and a loss of value, particularly as the company's success is closely tied to its human capital.

Therefore, it is crucial for entrepreneurs to anticipate potential challenges in securing capital and plan accordingly. One of the major pitfalls observed in biotech entrepreneurship is the tendency to start with insufficient capital, driven by a desire to avoid excessive dilution of ownership or a reluctance to engage in ongoing fundraising.

However, companies that commence operations with inadequate capital face a higher likelihood of failure, as the future financial landscape is unpredictable, even for seasoned investors like Warren Buffett. This issue will be further explored in the subsequent discussion.

Challenges for the Entrepreneur

In the context of entrepreneurial challenges, the primary concerns revolve around securing adequate funding to propel the company from its current valuation (V_t) to the subsequent valuation (V_{t+1})—essentially, achieving the next significant milestone. This necessitates a delicate balance: while entrepreneurs aim to secure sufficient capital, they also seek to avoid unnecessary dilution of their ownership stake. Additionally, they must prepare for potential financial disruptions by prudently planning for future capital needs, recognizing that capital markets may not always be readily accessible.

This process demands a measured approach, ensuring continual progress towards reaching the next value inflection point. Managing both the scientific and business aspects of the enterprise is crucial in this regard. Addressing these challenges involves careful consideration of various factors, including financing requirements, cost of capital, and program timelines. This can be particularly challenging for scientists and typical biotech entrepreneurs, who are often more inclined to focus on the scientific core of their enterprise, relegating financial and operational details to a secondary priority.

Collaboration or seeking expertise in these critical but less enjoyable aspects of business management is essential and should be prioritized early on. While these considerations may seem unfamiliar and enigmatic to those unaccustomed to concepts such as net present value and cost of capital, they are not inherently complex. Nonetheless, the lack of familiarity with these financial principles underscores the need for educational initiatives, such as our upcoming online course on healthcare finance, aimed at demystifying these concepts for aspiring entrepreneurs.

In addition to financial management, entrepreneurs must prioritize communication with existing and potential investors, recognizing that these stakeholders represent the primary source of capital. Demonstrating the ability to create value for shareholders at each value inflection point—V_0, V_1, $V2$—is crucial. This involves showcasing consistent growth in valuation, assuring investors that their investment is yielding positive returns.

Conversely, a decline in valuation between inflection points raises concerns and prompts questions about the company's performance and future prospects, potentially undermining investor confidence. By focusing on these key aspects, entrepreneurs can enhance their ability to attract and retain investor support, ultimately contributing to the sustained growth and success of their ventures.

Raising Capital

The process of raising capital in the biopharma industry has evolved significantly, offering a myriad of opportunities but also introducing greater complexity. Formerly, financial allocation and acquisition within this sector adhered to a specific model, as delineated in a Nature Biotechnology article composed by industry professionals. This model illustrated the various stages of drug development—comprising R&D, preclinical trials, and phases one through three, among others. In the earlier stages, funding for R&D and preclinical trials primarily emanated from familial and friendly sources, angel investors, and foundations due to the ambiguity surrounding potential returns in this realm, necessitating a substantial leap of faith and an early financial commitment.

As the process advanced into the preclinical and phase one stages, venture capitalists began to play a more pivotal role. Subsequently, as the drug's potential became increasingly evident during the clinical phases, a growing number of investors injected larger sums of capital into the enterprise. This progression was intrinsically linked to risk, with the highest level of risk concentrated at the ideation phase, characterized by considerable uncertainty regarding the commercial viability of a particular approach. Consequently, a limited number of individuals were willing to invest at this precarious juncture due to the unpredictability of potential returns.

As the research and clinical discovery advanced, the risk diminished with each successfully concluded phase. The most substantial reduction in risk occurred at the culmination of phase two, signaling a positive trajectory. Notably, a positive phase two outcome led to a significant surge in stock value, often prompting companies to go public. Furthermore, the approval of a drug substantially mitigated risk, rendering it akin to a secure asset. This traditional investment process was contingent upon the concept of de-risking, aimed at minimizing the risk for investors.

However, the contemporary financial landscape has ushered in a new paradigm, encompassing a diverse range of funding sources. Notably, family offices, foundations, venture philanthropy, patient advocacy groups, and even prominent pharmaceutical companies have established incubators and university collaborations, creating a multifaceted network of capital sources.

The modern era boasts numerous avenues for capital infusion, reflecting the inherent risks associated with clinical trials. This is encapsulated by the concept of the inverse funnel, representing the winnowing process from numerous drug candidates to a singular successful outcome, which is intrinsically linked to the Sharpe ratio.

The inverse of this ratio signifies the available investment capital. Remarkably, when an enterprise is perceived as a probable success, a multitude of investors are inclined to participate, underscoring the significance of investor confidence in the financial realm.

In the context of high levels of risk, only a small minority of investors are willing to take on such risks. Philanthropy has emerged as a valuable avenue for generating value with investors who comprehend the risks involved. Patient advocacy groups also serve as a significant source of funding, provided that their objectives align with those of the researchers seeking support. The Cystic Fibrosis Foundation, for instance, has established successful partnerships with investors and biotech entrepreneurs to develop treatments for CF patients.

Furthermore, pharmaceutical companies, biotech firms, and clinical research organizations (CROs) have increasingly shown interest in investing in emerging biotech ventures due to the potential impact on their business pipelines. Accessing these resources, however, poses a more significant challenge for individuals lacking business experience. Crafting a compelling narrative becomes crucial in persuading these organizations to invest. To aid in this endeavor, venture capitalist Bruce Boothe, a partner at Atlas Ventures, has outlined ten tips for raising funds in the biotech venture capital sector, which were published in 2013.

Boothe's tips emphasize the importance of understanding the investor audience, clarifying the nature of the venture, emphasizing the scientific merits, and assembling a team with a balance of experience and youthful vigor. It is essential to communicate the scientific potential of the project and to demonstrate passion for the research. Additionally, having a team with both experienced advisors and youthful enthusiasm is vital for attracting venture capital investment.

Few venture capitalists are willing to invest in entrepreneurs aged 50 or 60. The prevailing perspective is that individuals of this age who have achieved success should be able to fund their own ventures and startups. Although this may not always hold true for larger ventures, investors often seek youthful enthusiasm, considering it an advantage.

However, it is crucial for entrepreneurs to possess the necessary experience to navigate the challenges they will face. In the biotech industry, risk management is a key aspect of research and development. Entrepreneurs must have a deep understanding of the specific timelines, associated risks, and value inflection points for their projects.

Realistic representation of risk is essential, as venture capitalists are aware of the inherent overconfidence in entrepreneurship. Entrepreneurs are advised to ground their valuation expectations in reality, considering not only the current investors but also future ones. Overvaluing a company in the initial stages can lead to challenges in subsequent funding rounds, potentially damaging the company's reputation.

It is advisable to present a credible long-term plan for the company's capital needs and potential liquidity paths, demonstrating a thoughtful approach to managing finances. Additionally, establishing connections and seeking introductions from trusted sources are emphasized as crucial steps when engaging with venture capitalists, given the personalized and relationship-driven nature of the venture capital industry.

In the context of engaging with a venture capitalist, it is advisable to seek an introduction from a seasoned professional within the industry, as this approach is likely to result in a more earnest consideration of your business plan compared to a cold outreach. Such a method of introduction has proven to be significantly more successful than attempting to establish contact independently.

Moreover, it is imperative not to compromise on legal counsel, despite the associated costs. While it may be tempting to economize, especially after securing initial funding, it is essential to recognize the necessity of legal oversight to safeguard intellectual property. Although legal services may be expensive, their value, particularly in ensuring the proper protection of intellectual property, cannot be overstated.

Entrepreneurs should be prepared for the substantial fees that competent legal representation demands. Such fees are not indicative of exploitation, but rather reflect the standard rates prevalent in the market. It is crucial to establish a solid legal foundation from the outset.

Finally, entrepreneurs are encouraged to demonstrate unwavering passion and energy, traits highly valued by venture capitalists. VCs seek individuals who are not merely seeking funding for research, but rather aspire to effect significant change in the world, particularly in the current climate where the potential for transformative advancements in biomedicine is apparent.

New Business Models

What has been written on so far is the traditional biotech VC model. Here is an example of a company involved in bit of a different model:

A venture capitalist read a paper on rare disease therapeutics illustrating that if statistically low correlated projects are combined, the risk can be reduced rapidly, represented as 1 over the square root of n. They contacted the researcher and asked for the model to be run with a different set of assumptions to determine the probability of success and how much it would cost to go to phase one, phase two, and so on.

After a few months, the venture capitalist started a company to combine several rare disease therapeutics based on the researcher's simulations. They created a parent company, and the projects of different therapeutics were subsidiaries underneath that parent company.

One of the first investors into that company was a private equity fund. Private equity does not traditionally invest in VC funds. However, because the subsidiaries were structured as private companies, the capital received was a permanent part of the parent company which went on to raise over $500 million in multiple rounds, staged commitments and is valued at slightly over $1 billion.

The parent company has 18 rare disease therapeutics as subsidiaries, two of which have done IPOs. Their investors include private equity firms, an insurance company, hedge funds and biotech

VCs. This is different business model from the VC structure relying on the fact that they are a portfolio of rare diseases structured in a way to reduce the risk to investors.

Another example of a different business model is the Cystic Fibrosis Foundation with the initial goal of creating a drug for cystic fibrosis. In the 1990s, the CF Foundation started investing in pharmaceutical drug companies.

The foundation would engage with a biotechnology firm and express interest in collaborating to create a drug for cystic fibrosis, offering financial investment in exchange for equity in the company. However, instead of seeking equity, they pursued royalty interests in any approved drugs. This unique arrangement provided stability to the biotech companies, as the foundation's focus was on advancing the development of a cystic fibrosis drug rather than solely on financial returns.

The CF Foundation's sustained commitment to supporting biotech companies, such as Vertex, through financial investment and understanding during challenging periods, ultimately led to the approval of Kalydeco by the FDA in 2012. Subsequently, the foundation sold its royalty interests in Vertex for a substantial sum of $3.3 billion in 2014. This exemplifies the foundation's long-term perspective and dedication to advancing medical treatments for cystic fibrosis.

The significant financial gain from the sale has enabled the foundation to further its mission by pursuing gene therapy to ultimately cure the disease. This case serves as a compelling example of the influential role that patient advocacy groups, such as the CF Foundation, can play in advancing medical research and development.

Furthermore, the foundation's approach underscores the importance of maintaining a patient-centric focus in scientific and entrepreneurial endeavors, as it can attract necessary funding for innovative research and facilitate the translation of scientific discoveries into tangible clinical applications.

This case study highlights the potential of various business models, including drug royalty investment companies and securitization, to attract increased financial resources into the medical research and development space. The success of these approaches in reducing risk and attracting investor interest demonstrates the potential for leveraging finance to advance biomedical innovation.

8. SECURITIZATION

At this point in the textbook, a significant amount of material has been covered. The upcoming topic to be addressed is securitization, which is considered to be an influential tool for managing risk and generating substantial funding within the realm of healthcare finance.

Previous sections have highlighted the impact of financial risk on the availability of funding for advancing scientific breakthroughs from the laboratory to clinical application and widespread use. This chapter will focus on the application of securitization within the biomedical context. It is important to also address the financial crisis in relation to securitization, as improper implementation can lead to challenges, despite its potential to attract more capital for medical research.

The chapter will begin by examining the role of securitization in the financial crisis, followed by an exploration of its application to a diverse portfolio of biomedical projects. This examination will consider the functioning, size, and success rate of these projects, as well as the potential for failure.

The Financial Crisis, Financial Engineering & Securitization

One of the most successful pharmaceutical treatments in the history of oncology is pembrolizumab, a monoclonal antibody drug developed in 2014. It has proven to be highly effective, saving numerous lives despite its high cost. Consequently, there is a strong demand for this innovative drug, raising the question of how to finance its widespread use.

This chapter explores securitization as an alternative financing method in the biomedical industry, which offers distinct advantages compared to traditional financing approaches. Specifically, securitization can be utilized to establish large biomedical funds, known as mega funds, surpassing the typical scale of biomedical venture capital funds. To comprehend this concept, it is essential to examine the origins of the financial crisis, which can be traced back to the 1990s.

During this period, various financial institutions, including banks, mortgage companies, and government-sponsored enterprises, engaged in financial innovations related to mortgages and homeownership. These innovations led to the creation of numerous financial products associated with US residential real estate, attracting funding from a wide range of investors.

However, as interest rates began to rise around 2004, homeowners faced increasing difficulty in servicing their mortgages, resulting in a wave of defaults that disrupted the anticipated cash flows, ultimately leading to the collapse of several financial institutions. The aftermath of this crisis had far-reaching and devastating effects on individuals in the United States and worldwide.

The involvement of financial engineering and securitization played a significant role in the particular crisis under discussion. To illustrate this, consider a hypothetical scenario where a promissory note, represented by a green piece of paper, denotes an IOU for $1,000, payable at a future date to the holder.

This piece of paper can be likened to a bond or any fixed income obligation, albeit with an uncertain payment of $1,000, entailing an associated risk. Let's assume that it pays $1,000 with a probability of 90% and nothing with a probability of 10%, resulting in a 10% default rate, making it a risky bond.

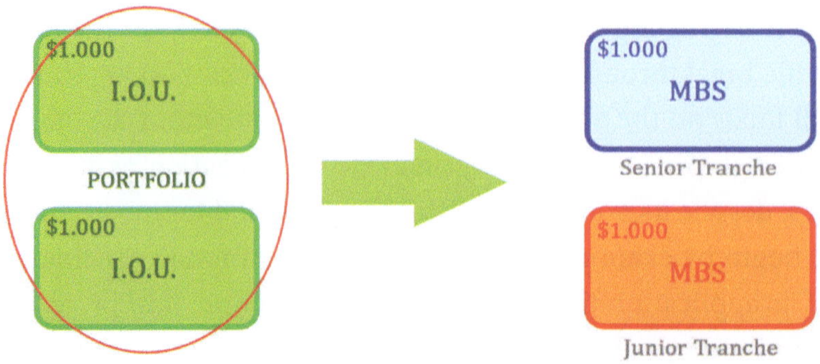

Figure 32: Promissory notes and fixed income obligations

Given the prevailing near-zero interest rates and its imminent maturity, how much would one be willing to pay for this piece of paper today? The anticipated value of the IOU is $900, which should represent its price. At $900, despite possessing a high default rate, it becomes a reasonably valued investment.

Essentially, the key point regarding the financial crisis and the so-called toxic assets is that while it may not initially appear to be an attractive bond, at a certain price, it becomes a viable investment. At $800, it transforms into a highly appealing asset. Therefore, the underlying premise is that there are no inherently bad bonds; rather, it is the pricing that determines their desirability.

Furthermore, considering the potential enhancement through portfolio theory, imagine combining two such bonds to create a portfolio. This involves establishing a legal entity, termed a special purpose vehicle, to own these two bonds and subsequently issuing new obligations based on them.

By doing so, it is possible to alter the risk profiles of the resulting instruments, which is contingent on the prioritization of payments. This introduces the concept of senior and junior tranches, where the former holds precedence in receiving payments over the latter. This alteration in priority fundamentally transforms the risk profile of these instruments, necessitating a mathematical analysis to comprehend their valuation.

It is assumed that each bond represents a statistically independent coin toss, with a 90% probability of repayment and a 10% probability of default. This single assumption encapsulates the essential characteristics of this specialized financial entity. Specifically, it is capable of assuming only three potential values: \$2,000, \$1,000, or nothing, with respective probabilities of 81%, 18%, and 1%.

Assuming Independent Defaults

Portfolio Value	Probability
\$2,000	81%
\$1,000	18%
\$0	1%

The 81% probability of both bonds paying off arises from the independent events of each bond paying off, which can be calculated by multiplying their individual probabilities. Conversely, the 1% probability represents the scenario where both bonds default. The remaining 18% probability pertains to the situation in which exactly one of the two bonds' defaults.

This probability distribution allows for the valuation of the distinct components within the financial structure. The prioritization of repayment is evident, as the blue bond must be paid \$1,000 before the orange bond. Consequently, in a scenario where the SPV holds a value of \$2,000, both the blue and orange bonds are honored, whereas in a \$1,000 scenario, only the blue bond is repaid. It is only when both green bonds default that the blue bondholders face a loss, an event with a likelihood of 1%. In contrast, the orange bonds carry a significantly higher default risk at 19%.

The financial model thus created presents the blue bonds as a relatively secure investment, with a mere 1% default probability, while the orange bonds are characterized by a much riskier 19% default probability. However, it is emphasized that the value of these bonds is contingent upon market equilibrium and the actions of astute traders. Consequently, the anticipated equilibrium prices for the blue and orange bonds are \$990 and \$810, respectively. The former, being nearly risk-free, commands a high price, while the latter, due to its substantial default risk, is valued significantly below par.

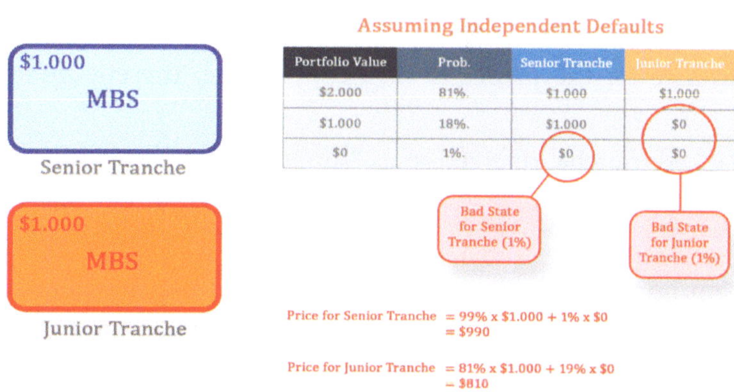

Figure 33: Assuming Independent Defaults

This transformation in the financial landscape results in the creation of two distinct markets, catering to different investor preferences. The blue bonds are likely to appeal to risk-averse entities such as pension funds, seeking secure investments. Conversely, the orange bonds are anticipated to attract hedge funds and other risk-seeking investors, given their inherently risky nature. By repackaging the initially unremarkable green bonds into these two new instruments, the financial entity has effectively broadened its appeal to diverse investor groups.

Moreover, this transformation also facilitates transparency and information availability for rating agencies, enabling them to provide informed assessments of the value of these securities. A hypothetical scenario can be considered where you assume the role of a rating agency tasked with determining the appropriate rating for these bonds in question.

The assessment of the bonds is contingent upon factors such as their historical default rates and the established correlation between ratings and defaults. An examination of historical data from Moody's spanning the years 1920 to 2008 reveals the realized default rates across various bond ratings after one, five, 10, and 20 years. Each rating category is represented by four bars, each denoting the default rates at different time intervals following the issuance of the bonds.

For instance, the data indicates that 10 years after the issuance of AAA-rated bonds, the default rate stands at 0.9 percent. Conversely, bonds rated as BA exhibit a 10-year default rate of 18.4 percent. Based on these statistics, one can infer the appropriate rating for the bonds in question. Notably, bonds with a 1 percent default rate would be categorized as AAA, while those with an 18.4 percent default rate would be classified as BA.

Subsequently, the rating agencies utilize this quantitative analysis and information provided by the issuers to assign ratings to the bonds. It is observed that the blue bonds are rated AAA, while the orange bonds are rated BA, signifying that they fall below investment grade. Consequently, certain financial institutions, such as pension funds, money market funds, and commercial banks, are prohibited from investing in BA-rated bonds, with the exception of hedge funds.

Furthermore, even bonds with a AAA rating may not be considered suitable for investment by certain financial institutions, such as money market funds, which prioritize safeguarding against any potential loss. In response to this concern, the issuers devised a strategy to mitigate principal risk by offering insurance on the principal at their expense, thereby guaranteeing protection against default.

This insurance-backed tranche, referred to as "super senior," was positioned as a more secure investment than AAA-rated bonds, as it incorporated an additional layer of insurance from highly-rated insurers, such as AIG. Consequently, a substantial volume of this paper was issued, attracting interest from hedge funds due to the potential for a significant rate of return, despite the associated risks. This led to a substantial issuance of the paper, highlighting the magnitude of this market.

In 1996, an analysis of the total mortgage-related debt issued during that year in the United States revealed a sum of approximately $480 billion. Fast forward to 2003, and this figure had surged to around $3 trillion, representing a substantial increase in mortgage-related debt issuance. These financial instruments were designed to be highly appealing to a wide range of investors, leading to varying perspectives on their attractiveness. The market's appeal stemmed not only from differing opinions but also from the distinct risk aversion parameters held by different stakeholders, influencing their willingness to assume risk. While agreement on the level of risk might exist, disparities emerged in the inclination to embrace such risk.

The surge in financial innovation and the proliferation of these financial instruments precipitated a housing market boom. This was evidenced by the Shiller index of US residential real estate, which depicted a notable upswing from 1995 to 2006. The period saw a near doubling of US residential real estate values, with soaring home prices fostering a climate of optimism and encouraging individuals to purchase homes, often leveraging mortgages and relying on subsequent refinancing.

However, this upward trajectory was not sustainable. The economic downturn of 2008 shed light on the unforeseen complexities underlying the market. A chief risk officer's account in an article published by The Economist in 2008 revealed unexpected fluctuations in the value of different tranches of mortgage-backed securities, leading to losses and highlighting the intricate interplay of market forces.

The narrative underscored the challenges of comprehending and managing the inherent correlations between various financial instruments, ultimately revealing the need for a more nuanced modeling of these dynamics. The events described serve as a case study in the complexities of financial markets and the imperative of understanding and modeling the intricate interdependencies that underpin them.

The Influence of Correlation & their Limitations

Returning to the blue and orange pieces of paper., assuming the independent nature of the underlying green bonds, the probability table takes on a specific form. In the absence of any correlation among defaults, the pricing stands at $990 for the blue bond and $810 for the orange bond. However, if defaults become highly correlated, or even perfectly correlated, a shift occurs. This change is rooted in the assumption of independence and the potential for temporal shifts, as illustrated by the accompanying graph.

During upward market trends, such as the nationwide rise in US residential real estate, default rates across different localities are seemingly unrelated. However, when markets decline, a different narrative emerges. The correlation across localities becomes more pronounced, leading to a significant impact on the financial landscape. In the event of perfect correlation among the green bonds, the potential outcomes narrow down to two possibilities: both bonds either default or pay. Consequently, the values of these bonds converge to $900, rendering the priority inconsequential. This scenario presents a substantial challenge for AAA tranche holders and money-market funds, as the devaluation of their investments can be particularly detrimental.

The ensuing repercussions are far-reaching, as evidenced by the strain on insurance companies and the subsequent failure of insurers. Notably, hedge funds emerge as beneficiaries due to the altered correlations, highlighting the profound impact of this shift. The elusive nature of correlation as a parameter further complicates the situation, as it is not readily available for reference in standard financial publications. Even with risk managers' attempts to model correlations based on historical mortgage data, the unprecedented downturn in the US residential real estate market catches the banking sector off guard.

This sequence of events leads to the gradual erosion of equity and the eventual freeze of markets, culminating in the financial crisis. The crisis, with its enduring global reverberations, serves as a stark reminder of the perils associated with overlooking or underestimating correlated risks. The lingering aftermath of this crisis underscores the need for caution and foresight, particularly in the context of future endeavors, such as biomedical innovation.

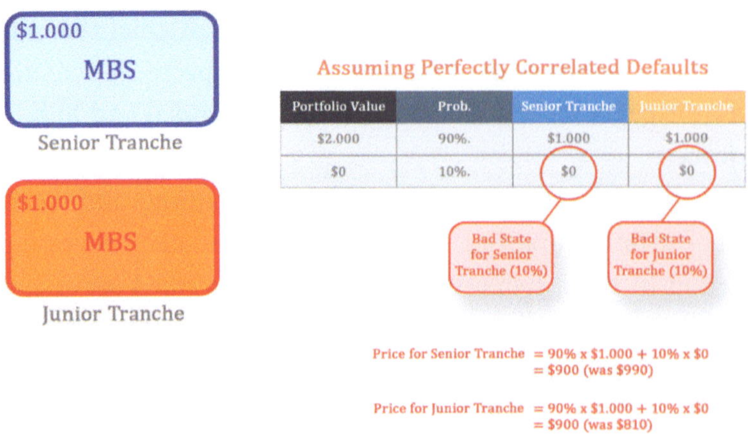

Figure 34: Assuming Perfectly Correlated Defaults

The application of the concepts discussed thus far is relatively straightforward. The inquiry posed is whether the same methodologies can be applied to finance cancer drug development projects, with the aim of expanding the resources available to the industry. The subsequent investigation revealed that conducting such a simulation was not overly complex, despite the challenges encountered in obtaining and calibrating the necessary data. The structure of the model is, in essence, uncomplicated. It involves a range of financial obligations—comprising senior debt, junior debt, and equity—that are utilized to fund various assets.

While this approach differs from traditional accounting practices, it serves to illustrate a natural progression, commencing with investors who provide funding in exchange for financial instruments. These funds are then utilized to acquire a portfolio of assets, predominantly comprising intellectual property, such as royalty interests or biomedical projects licensed from universities. These assets undergo clinical trials, with the ultimate goal of yielding approved drugs and generating revenue through sales or licensing. This revenue stream is then allocated to satisfy the diverse claims on the assets.

The central concept revolves around a mega fund, dedicated specifically to financing biomedical innovation, which has not conventionally been utilized for this purpose. The simulation of such a structure can be achieved through a straightforward approach. The model considers the progression of a preclinical asset through various developmental phases, with the potential for success or failure at each stage. The proposal entails managing a multitude of such projects simultaneously within a dedicated investment vehicle, thereby increasing the likelihood of generating cash flows. The financial focus is on determining the potential cash generation over a specific timeframe.

The subsequent step involves structuring the financing tools and capital arrangements, employing financial engineering techniques to prioritize the claims of investors on the cash flows. This may involve issuing triple-A bonds as the primary claimants, followed by other tiers of stakeholders, culminating in equity holders. This approach, while it may appear intricate, is not entirely novel, having been employed in various sectors such as mortgages, auto loans, and royalties. However, its application in biomedicine is an emerging trend, as described in the subsequent sections.

Modeling Correlations

In order to provide an understanding of the data utilized and to structure a large investment fund to capture these varying dynamics within a dataset of cancer compounds, the following information is presented. From a statistical standpoint, a diverse array of cancer compounds exists. Cancer, as a disease, encompasses numerous distinct conditions.

The comprehensive database of diversification contained 733 anticancer compounds, and the subsequent statistics pertain to these 733 compounds. Utilizing this data and its associated statistics to calibrate simulations, the probabilities of success exhibited considerable variation. Two simulations were conducted: the first simulation encompassed the period from preclinical to phase two (the early stages), while the second simulation spanned from phase two to NDA.

This division was recommended by industry experts, who highlighted that contemporary drug development typically involves biotech and venture capitalists in the early stages, while pharmaceutical companies are involved in the later stages. By the time a drug reaches phase 2b approval, it is typically sold to a pharmaceutical company, which then oversees its progression from phase 2b to NDA. The results of the simulations were particularly striking. For the early-stage fund, valued at approximately $5 billion, the equity tranche yielded a return rate of approximately 8.9%.

Additionally, the debt tranches, assumed to yield 5% and 8%, were considered quite generous, surpassing market rates for bonds of equivalent risk at that time. In contrast, the later stage fund, valued at $15 billion, yielded an equity return rate of 11.4%, with bond risks aligning with industry-appropriate ratings categories at 9% and 11%.

While these returns may not be considered exceptional from a venture capital perspective, they are regarded as favorable by entities such as pension funds, mutual funds, and retail investors, particularly when considering the substantial scale of the capital pools involved.

Nevertheless, several elements remain incompletely understood, with one common challenge revolving around the determination of the appropriate fund size. The concise response to this issue is that the optimal size is contingent upon a multitude of parameters derived from both academic literature and industry expertise.

Denote Success/Failure of Project i by $I_i = \{0,1\}$

$$\text{Payoff} = (I_1 + I_2 + \cdots + I_n) \times \text{rNPV} \tag{8.1}$$

$$\Pr\left(\sum_{i=1}^{n} I_i = k\right) = \sum_{j=0}^{k} \binom{n}{j} p^j (1-p)^{n-j} \tag{5.23}$$

$$\Pr\left(\sum_{i=1}^{n} I_i = k\right) = \sum_{j=0}^{k} \binom{n}{j} p^j (1-p)^{n-j} \tag{5.23}$$

$$I_i = \begin{cases} 1 & \text{if } X_i > \gamma_i \\ 0 & \text{if } X_i \leq \gamma_i \end{cases}, \quad X_i \sim \mathcal{N}(\mu_i, \sigma_i^2) \quad \begin{bmatrix} X_1 \\ X_2 \\ \vdots \\ X_n \end{bmatrix} \sim \mathcal{N}(\mu, \Sigma) \tag{5.24}$$

Ultimately, this underscores the necessity for collaboration between the financial and biomedical sectors. One critical aspect involves the consideration of correlation. Concerns regarding correlation can be addressed through the development of a specific mathematical model that begins with the evaluation of success or failure. This involves the use of an indicator variable, denoted as , which signifies the success or failure of an individual drug development project.

A value of one indicates success, while zero indicates failure. The fundamental concept underlying a comprehensive simulation is essentially a tally of successes, multiplied by a risk-adjusted NPV. While this approach can be refined to encompass various projects and cancer types, as well as diverse patient populations, the fundamental calculation remains unchanged.

Consequently, the key requirement is to ascertain the distribution of the sum of successes. If these outcomes are independent and identically distributed, akin to the two green pieces of paper initially considered, the resulting distribution is straightforward and conforms to a binomial distribution. However, in scenarios where correlation is not absent, alternative methods are necessary to account for this. Drawing from experiences such as those observed in the housing market, it is evident that correlations are not invariably zero.

Therefore, it becomes imperative to address this consideration. One potential approach involves the introduction of a continuous unobserved variable, denoted as , assumed to adhere to a normal distribution. This variable is introduced for each drug development project, with the assertion that a specific threshold, denoted as , determines project success or failure.

This continuous variable, when compared against , dictates the probability of success versus failure. Consequently, this introduces the notion of an unobserved continuous variable that governs project outcomes. However, the model becomes truly compelling when this continuous variable is assumed to be correlated across projects, thereby introducing correlation into the model.

By assuming a multitude of these indicator variable drivers and their correlation via a covariance matrix denoted as sigma, this correlation is effectively integrated into the model. While this is not the sole approach, it represents a particularly convenient method of addressing this consideration. The covariance matrix, sigma, aligns with the familiar concept from standard portfolio theory, albeit in this context, it pertains to drug development project outcomes rather than rates of return.

Typically, the covariance matrix encompasses variances along the diagonal and covariances along the off-diagonal elements. It is often simpler to consider correlations, and these can be derived by pre- and post-multiplying by the standard deviations. The result is a correlation matrix, which encapsulates the intuition of correlations as values ranging from -1 to 1, signifying perfect anti-correlation to perfect correlation, with zero indicating no relationship.

Consequently, the primary focus involves modeling this correlation matrix, thereby determining whether the failure of one project imparts any information regarding the likelihood of another project's failure. This necessitates a comprehensive approach to modeling correlation, either explicitly or implicitly, within this domain.

$$\Sigma = \begin{bmatrix} \sigma_1^2 & \sigma_{12} & \cdots & \sigma_{1n} \\ \sigma_{21} & \sigma_2^2 & \cdots & \sigma_{2n} \\ \vdots & \vdots & \ddots & \vdots \\ \sigma_{n1} & \sigma_{n2} & \cdots & \sigma_n^2 \end{bmatrix}$$

$$= \begin{bmatrix} \sigma_1 & 0 & \cdots & 0 \\ 0 & \sigma_2 & \cdots & 0 \\ \vdots & \vdots & \ddots & \vdots \\ 0 & 0 & \cdots & \sigma_n \end{bmatrix} \begin{bmatrix} 1 & \rho_{12} & \cdots & \rho_{1n} \\ \rho_{21} & 1 & \cdots & \rho_{2n} \\ \vdots & \vdots & \ddots & \vdots \\ \rho_{n1} & \rho_{n2} & \cdots & 1 \end{bmatrix} \begin{bmatrix} \sigma_1 & 0 & \cdots & 0 \\ 0 & \sigma_2 & \cdots & 0 \\ \vdots & \vdots & \ddots & \vdots \\ 0 & 0 & \cdots & \sigma_n \end{bmatrix} \quad (5.25)$$

In the case of early-stage assets with no cash flows, two options are available. One option is to issue more debt than necessary and use some of that debt to cover interest payments in the initial years. This strategy is commonly used in commercial real estate development, where cash flows are not realized until the property is completed and occupied.

The excess debt is disclosed to bond holders, who appreciate the regular cash flows it provides. Another approach is to initially rely on equity and gradually increase leverage as cash flows become available. This method, known as dynamic leverage, was detailed in a recent academic paper and is a common practice in other industries.

Turning to the challenges that arise when these aforementioned methods prove ineffective, particularly in the context of Alzheimer's disease, the allocated $30 billion may not suffice for a dedicated Alzheimer's fund due to the disease's formidable nature, making drug development and treatment exceptionally arduous. Developing an Alzheimer's drug typically spans 13 years, necessitating early detection and larger clinical trials, resulting in out-of-pocket costs ranging from $500 to $600 million.

The likelihood of success is markedly below 5%, reflecting the limited scope for progress in this domain. Unlike cancer research, there is a dearth of scientific understanding of Alzheimer's biology, hindering diversification efforts even if adequate funding were available. Consequently, the private sector is disinclined to establish an Alzheimer's-focused fund.

A review paper from 2014 underscored the daunting and faltering nature of Alzheimer's drug development, revealing a 99.6% failure rate based on nearly 400 trials conducted over an 11-year period. Notably, only one trial achieved success, albeit in 2003, with the drug Namenda addressing a symptom, not the disease itself. The high failure rate is further evidenced by recent setbacks, highlighting the urgency for a more comprehensive approach. Pursuing parallel endeavors is advocated to counteract the protracted waiting period for successful drug development, which, at a 1% success rate, could extend to 1,300 years. The associated costs are substantial, with projected 2017 Medicare and Medicaid expenses estimated at $175 billion solely for Alzheimer's treatment, excluding care costs and lost GDP revenues.

It is well known that correlation matrices can exhibit peculiar behaviors when populated with random numbers, potentially leading to non-compliance with the standards of a legitimate correlation matrix. This can manifest in the form of negative portfolio variances, which is an undesirable outcome given the impossibility of negative variances.

However, the practical application of such methodologies faces significant challenges, especially in the context of investment in an Alzheimer's *Megafund*. Despite the potential long-term benefits and cost savings associated with addressing Alzheimer's, private sector investors are unlikely to engage in such ventures, necessitating government support.

It's important to recognize the simultaneous progress being made in scientific research. Breakthroughs in high throughput screening for anti-amyloid plaque inhibitors, advancements in overcoming the blood-brain barrier for drug delivery, and the promising potential for reversing amyloid plaques are all indicative of significant strides in the field. Yet, the lack of adequate funding and the need to alter the risk-reward trade-off remain as substantial barriers to further progress.

Beyond Alzheimer's, similar challenges are prevalent in other critical domains such as pediatric oncology, vaccine development, clean energy, climate change, asteroid mining, and space colonization. These areas often represent long-shot endeavors, demanding innovative funding mechanisms and public-private partnerships to drive progress.

In light of these considerations, it becomes imperative to explore the feasibility and economic viability of funding such long-shot endeavors. While substantial amounts of capital are indeed required, recent examples, such as the $100 billion technology fund partnership between SoftBank and the government of Saudi Arabia, illustrate the availability of significant financial resources. The key question, however, revolves around the ability to generate meaningful returns on these investments, as evidenced by the underlying motivations of investors and funders.

As this field moves forward, it is essential to delve into the potential avenues for funding long-shot endeavors, both from a generic perspective and within the context of specific business models. By addressing these questions, we can better understand the economic landscape and the potential for achieving impactful outcomes in these critical areas.

9. CLINICAL TRIALS

The discussion thus far has emphasized the importance of considering the risk associated with a project's valuation, particularly in the context of the trade-off between risk and return. Given the significant implications of risk in biomedicine, particularly in cases involving life and death, careful consideration of risk is essential.

This chapter takes a more rigorous and direct approach to studying risk, moving away from abstract financial concepts and focusing on the intricacies of the drug approval process. This includes an analysis of the statistical perspective on risks. Specifically, the chapter will examine the process of conducting a typical randomized clinical trial and the evaluation of results to determine the approval of a new drug or device.

The statistical properties of clinical trials, such as size, power, and cost, will be discussed in relation to the financial valuation of a biomedical project. This statistical analysis is likely to be the most challenging aspect of the course, requiring a thorough understanding. However, it is also crucial in understanding how risk is managed in the drug development process. Furthermore, these statistics have significant financial implications, underscoring the need for a detailed examination of their technical aspects.

An Overview on Drug Development & Clinical Trials

With biologics assuming a more prominent role in certain therapeutic approaches, the expenses associated with drug production have escalated significantly, particularly during the clinical trial phase. This financial burden, which extends beyond the trial phase to commercial production after approval, has become more complex.

Consequently, this challenge has a direct impact on the process of securing funds and conducting the necessary experiments and trials. It is imperative to gain an overview of the drug development process and the associated analytics.

The initial phase of this process centers around the patient, a critical aspect often overshadowed by the predominant focus on business aspects in educational settings. It is vital to emphasize that drug development fundamentally revolves around patients, and this perspective should not be disregarded. This prompts the need to question the underlying reasons for the time-consuming nature, high costs, and meticulous control of these experiments.

The drug development process commences in the laboratory, where diverse compounds or devices with potential therapeutic properties are developed, necessitating extensive research to comprehend the fundamental science behind these potential therapies. Subsequently, these therapies are tested in controlled settings, typically involving animal models such as mice or dogs.

Based on the outcomes of these animal models and other studies, the transition to human trials becomes a pivotal and highly regulated step. Prior to any clinical experimentation, an Investigational New Drug (IND) application must be compiled and submitted to the FDA for review and approval. The FDA's extensive repository of information underscores the significance of heeding their advice. Once the IND is approved, the drug can progress to clinical trials, which encompass three distinct phases: phase one, phase two, and phase three.

Phase one, typically involving a small study group of 50 to 100 individuals, primarily focuses on determining safe dosages, with efficacy being a secondary consideration. Phase two, encompassing a larger group of 100 to 500 participants, shifts the focus to biological activity and potential therapeutic effects. Should the drug exhibit promise in phase two, it advances to phase three, involving a significantly larger group of 1,000 to 5,000 individuals, with the primary objective of assessing clinical effectiveness.

Each phase involves a gradual increase in the number of participants to ensure a thorough understanding of the safety profile of the drug or device. Subsequent to successful completion of phase three, an application known as an NDA is filed, and the FDA undertakes a rigorous review process, which includes scrutiny of all relevant data and records from clinical trials.

If the application is approved, the drug can then be marketed. Additionally, in some cases, a phase four study may be mandated by the FDA to monitor patient responses following the commencement of drug marketing and widespread usage.

An examination of the quantitative aspects of clinical trials provides an understanding of the substantial scale of these endeavors. In 2015, the pharmaceutical industry's trade organization, PHRMA, in conjunction with Battelle, released a report detailing the current status of clinical trials, offering insight into the volume of ongoing trials at any given time. In 2013, approximately one million individuals participated in clinical trials, spanning phases one, two, and three.

While this figure is considerable, it is also relatively small when viewed from a different perspective. Notably, within the realm of cancer alone, an estimated 14 million Americans are afflicted with various forms of the disease. Surprisingly, it is estimated that only 5% of clinical trials in cancer involve cancer patients, a disproportionately low number considering the potential impact and knowledge that could be gained from increased participation.

The data reveals that the largest number of trials occurs within the domain of cancer, given the diverse types of cancer and the numerous products currently under development for its treatment. However, the fraction of cancer patients participating in trials remains notably minuscule. One of the primary challenges confronting the industry is the issue of accrual—how to attract more patients to participate in these trials. The greater the number of participants, the more comprehensive the understanding of the safety and efficacy of drugs.

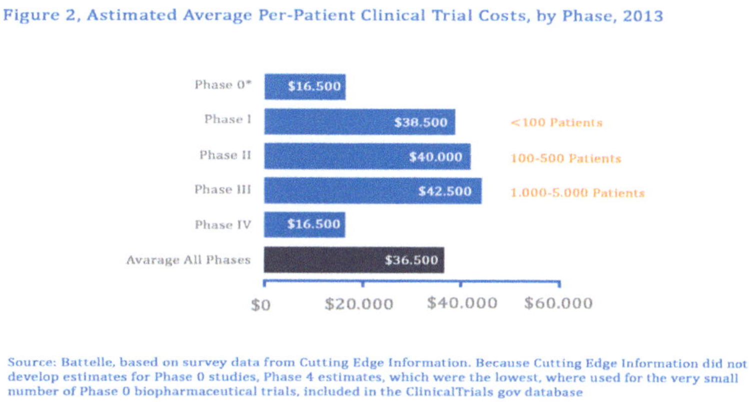

Figure 35: Average Patient Clinical Trial Costs by Phase (Source: Cutting Edge Information)

The discussion then shifts to the financial aspect, as it directly influences the analytical assessments. Clinical trials are notably expensive endeavors, with costs amounting to approximately $36,000 per patient per phase, and roughly double that amount for cancer trials.

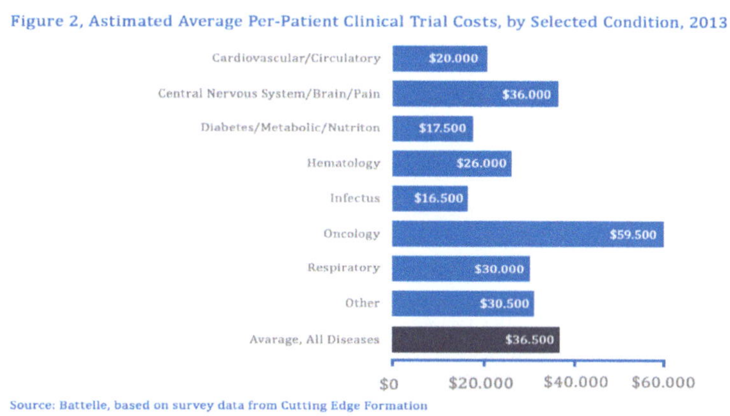

Figure 36: Average Patient Clinical Trial Costs by Selected Condition (Source: Cutting Edge Information)

When these figures are multiplied by the number of patients, the substantial financial commitment becomes evident, with costs ranging from five to ten million in phase one and 50 to 100 million in phase two. Furthermore, applicants are required to pay the FDA for the application process, as mandated by the Patient Drug User Fee Act of 1992, which has significantly reduced the median review time for applications.

In 2017, the application for a new drug cost approximately $2 million, a seemingly substantial sum that, upon closer examination, reflects the extensive review and inspection processes undertaken by the FDA. The complexity and increased expenses associated with clinical trials are attributed to the growing intricacy of underlying disease biology, as evidenced by examples of escalating complexity over the past decade. This complexity contributes to the escalating costs and challenges associated with clinical trials, a phenomenon exemplified by Eroom's Law, which asserts that these endeavors are becoming increasingly arduous and costly.

Randomized Clinical Trials

In the context of a randomized clinical trial design and its impact on investment risk, the fundamental concept revolves around conducting a controlled experiment to evaluate a specific therapy on a patient group. This entails the inclusion of a control group to discern whether the outcomes are influenced by the placebo effect or the actual therapy.

The emphasis is on prospective studies, as opposed to retrospective ones, to avoid biased interpretations of historical data and focus on future outcomes. The design aims to steer clear of post-hoc justifications and instead proactively study future events without prior knowledge of the results, thereby avoiding any potential manipulation of the study.

This approach is considered the gold standard in biomedicine analysis due to its ability to mitigate back test bias and overfitting. The trial typically involves dividing a patient sample into two groups, one receiving the treatment and the other a placebo or standard care.

At the trial's conclusion, a comparison is made between the outcomes of the two groups, with the expectation that the treatment group will exhibit superior results based on predetermined metrics. The approval of a treatment depends on the trial's purpose, such as non-inferiority or toxicity assessment, and should be pre-specified rather than retroactively claimed.

The primary rationale for conducting a clinical trial and its recognition as the benchmark standard is rooted in the concept of predisposed biases, data exploration, and exhaustive data analysis until desired outcomes are achieved. The fundamental purpose of a randomized clinical trial is to mitigate the likelihood of false positives, rather than solely to combat fraudulent activities, which can occur in any industry, including the field of medicine. Well-intentioned, ethical researchers may inadvertently deceive themselves due to their strong desire to believe in the efficacy of their drug, particularly when patient well-being is at stake.

Emotional investment in the outcomes of clinical trials can lead to cognitive biases, causing individuals to justify results post hoc, even when patterns are non-existent. People are inclined to identify patterns and are drawn to the unusual, which can influence their assessments of likelihood. Even individuals with good intentions are susceptible to this cognitive bias.

A study conducted a few years ago revealed that nearly half of the psychological studies analyzed in top-tier journals could not be replicated, indicating that our perceptions subtly influence conclusions. The imperative of a randomized clinical trial is to prevent such biases from impacting results, as the approval of an ineffective drug can result in fatalities from adverse effects.

Regulatory bodies such as the FDA mandate specific requirements for clinical trials to safeguard against biased outcomes. Central to these requirements are the establishment of predefined endpoints, which serve as the basis for evaluation. Additionally, patient selection and accrual are critical considerations, as biased patient selection can compromise the integrity of the study.

Randomization of patient selection and assignment to treatment or control groups is therefore essential, and while more costly, it at least mitigates the risk of skewed results. Representative sampling and biomarker studies are now indispensable in ensuring the generalizability of trial outcomes, adding complexity and expense to the process.

The Design and Ethics of Clinical Trials

The process of designing clinical trials is a multifaceted endeavor that demands careful consideration of various methodologies and their implications on patient outcomes, research costs, and regulatory approval. One of the fundamental aspects of trial design is the choice between parallel and crossover designs.

The parallel design, a traditional approach, involves the segregation of patients into distinct control and treatment groups, which are followed throughout the trial and subsequently compared. In contrast, the crossover design, a more recent and somewhat controversial approach, utilizes a single pool of patients for both control and treatment purposes. This involves administering the treatment to one group while providing a placebo to the other, and then switching the treatments after a certain period for comparative analysis.

The crossover design presents two primary attractions. Firstly, it appears to offer cost savings by potentially requiring a smaller patient pool and reduced follow-up periods. Secondly, and perhaps more significantly, it addresses ethical concerns regarding patient participation. In the context of clinical trials, patients facing serious illnesses are often reluctant to be assigned to a placebo group, as they view the trial as their last hope for treatment. The crossover design offers the reassurance that all patients will eventually receive the treatment, albeit at different times, thus alleviating the psychological burden associated with the possibility of receiving a placebo.

However, the adoption of the crossover design is not without its controversies and risks. While it addresses ethical concerns, there are concerns about potential biases and confounding factors that may arise from the crossover approach. Furthermore, the regulatory bodies such as the FDA require robust evidence of clinical efficacy, and the design of the trial plays a pivotal role in determining the approval or rejection of a drug candidate.

It is important to acknowledge that while there are instances where clinical trials do not adhere to the traditional two-arm, randomized, balanced, controlled trial structure, the presence of a control group is imperative for ensuring the validity and reliability of the study outcomes. The absence of a control group can lead to challenges in demonstrating clinical efficacy, potentially resulting in negative regulatory decisions that have far-reaching implications for all stakeholders involved. It is essential to recognize the pivotal role trial design plays in shaping the outcomes of drug development and the subsequent impact on patient care, investor confidence, and the advancement of medical science.

It is now evident that the crossover trial design constitutes a method wherein patients initially assigned to a control group are allowed to switch to the treatment group at a later stage. This approach has gained traction among clinicians due to its perceived ethical implications, ensuring that all participants eventually receive the experimental drug. However, the decision-making process regarding when patients are permitted to crossover is often linked to the severity of the disease. Consequently, this selection mechanism tends to favor sicker patients, potentially impacting the overall survival and efficacy outcomes. The ethical complexities of crossover trials are further compounded by the need to maintain clinical equipoise, wherein the uncertainty of a drug's effectiveness justifies the presence of both control and treatment groups.

Blinding, a crucial aspect of trial design, aims to prevent biases in outcome assessment. Double-blind studies, where neither the patients nor the investigators are aware of group assignments, are preferred, but in certain cases, such as surgical interventions, single-blinded studies may be the only feasible option. The reporting of trial results adheres to standardized guidelines to ensure comparability across different studies.

While crossover practices in clinical trials can yield valuable data and address ethical concerns, they also carry potential drawbacks if not implemented judiciously. The integrity of trial data must be maintained, and decisions regarding trial design and execution should not be driven solely by cost considerations. The interplay between trial conduct and funding underscores the critical relationship between these factors, as evidenced by instances where cost-saving measures influenced the adoption of crossover designs in Eastern Europe, ultimately impacting the outcomes of the trials.

Statistics and Clinical Trials

The statistical aspect of a clinical trial is an essential component of research. A balanced two-arm randomized clinical trial is a common design for such studies. In this design, there are two groups of n patients each - a control group and a treatment group. The drug being tested is assumed to have a continuous effect that can be quantitatively measured, such as the reduction of blood pressure or the shrinkage of tumors. The focus here is on continuous variables, which simplifies the statistical analysis.

Key Parameters

α = Type I error

β = Type II error , $1-\beta$ = Power

n = Sample Size for Each Arm

$X_i \sim \mathcal{N}(\mu_x, \sigma^2)$ = Patient i's Response, Control Arm

$T_i \sim \mathcal{N}(\mu_t, \sigma^2)$ = Patient i's Response, Treatment Arm

$\delta = \mu_t - \mu_t$ = Expected Treatment Effect Size

σ^2 = Variance of Patient Outcome

H_0 = Null Hypothesis, $\delta = 0$

H_1 = Alternative Hypothesis, $\delta = \kappa\sigma > 0$

The primary objective of the clinical trial is to determine whether the mean response rate in the treatment group (μ_t) is equal to the mean response rate in the control group (μ_x). The null hypothesis posits that there is no effect, while the alternative hypothesis suggests that the drug is effective. The statistical evaluation of these hypotheses involves constructing confidence intervals and calculating estimates of the difference in response rates. If the estimated difference falls outside the confidence interval, the null hypothesis is rejected in favor of the alternative hypothesis.

$$T_i \overset{\text{IID}}{\sim} \mathcal{N}(\mu_t, \sigma^2) \quad , \quad X_i \overset{\text{IID}}{\sim} \mathcal{N}(\mu_x, \sigma^2) \quad , \quad i = 1, \cdots, n$$

Where:

Null hypothesis H_0: $\delta \equiv \mu_t - \mu_x = 0$ (no effect)

Alternative hypothesis H_1: $\delta > 0$ (positive effect)

$$\hat{\delta} \equiv \hat{\mu}_t - \hat{\mu}_x = \frac{1}{n}\sum_{i=1}^{n} T_i - \frac{1}{n}\sum_{i=1}^{n} X_i \tag{9.1}$$

The process of evaluating a drug's effectiveness relies heavily on statistical analysis. By understanding the statistics of the clinical trial, researchers can effectively manage the risk associated with drug development. Hypothesis testing and the construction of confidence intervals are fundamental tools in this process, allowing researchers to make informed decisions about the effectiveness of the drug being studied.

- $\hat{\delta}$ is a random variable; its distribution is needed to determine whether data is consistent with null or alternative
- $Z_n \equiv (\hat{\delta} - \mathrm{E}[\hat{\delta}])/\mathrm{SD}[\hat{\delta}] = \hat{\delta}/\mathrm{SD}[\hat{\delta}] \sim \mathcal{N}(0,1)$ under H_0 (9.2)
- Decision rule:
  - if $Z_n \leq c$, accept H_0 (reject drug)
  - if $Z_n > c$, accept H_1 (approve drug)

The use of a 95% confidence interval and a critical threshold of 1.96 for a two-sided confidence band are common practices in this context. These statistical methods help to minimize the risk of making false positive conclusions about the effectiveness of the drug under study.

- Under null hypothesis H0: $\delta \equiv \mu_t - \mu_x = 0$ (no effect):

$\alpha = \text{Prob}(-c < \hat{\delta} < c)$ (α% Confidence Interval)

$$\frac{\alpha}{2} = \text{Prob}(\hat{\delta} > c) = \text{Prob}\left(\frac{\hat{\delta}}{\sqrt{\text{Var}[\hat{\delta}]}} > \frac{c}{\sqrt{\text{Var}[\hat{\delta}]}}\right)$$

$$= \text{Prob}\left(Z > \frac{c}{\sigma\sqrt{2/n}}\right) = 1 - \text{Prob}\left(Z \leq \frac{c}{\sigma\sqrt{2/n}}\right)$$

$$\frac{\alpha}{2} = 1 - \Phi\left(\frac{c}{\sigma\sqrt{2/n}}\right) \Rightarrow \frac{c}{\sigma\sqrt{2/n}} = \Phi^{-1}\left(1 - \frac{\alpha}{2}\right) \equiv z_{\alpha/2}$$

$$c = \sigma z_{\alpha/2}\sqrt{2/n} \qquad (9.3)$$

For $\alpha = 5\%, z_{\alpha/2} = 1.96$

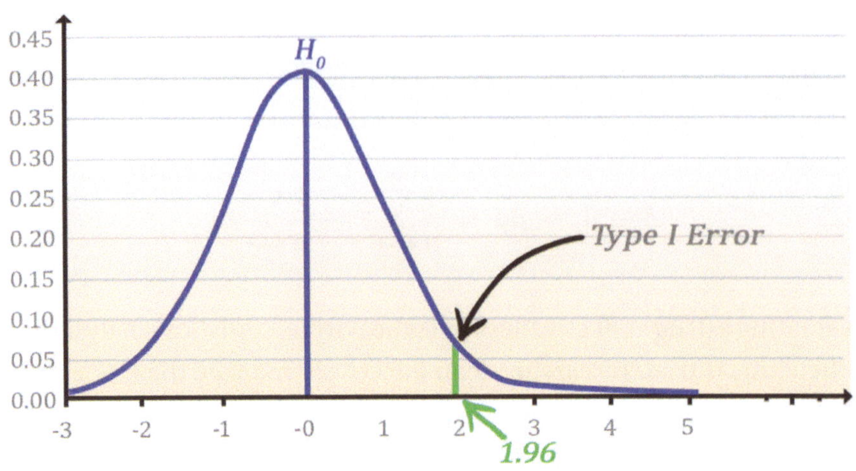

Figure 37: Type I error

In the course of these computations, one must consider a pivotal choice: whether to grant approval or to decline it. The critical inquiry pertains to the ramifications of this decision. There exist two scenarios: one where an ineffective medication is dismissed, and another where an effective medication is sanctioned. Consequently, two potential errors emerge. Firstly, a false positive occurs when a drug lacking efficacy is approved erroneously, denoted by the probability α in a two-sided test.

Secondly, there is the risk of committing a Type II error, which involves failing to approve a genuinely efficacious drug due to insufficient supportive data. This trade-off between Type I and Type II errors is a focal point for regulatory agencies and scientific communities, aiming to minimize the occurrence of both errors. While it is theoretically feasible to eliminate either type of error entirely, the practicality and desirability of such extremes are questionable.

To eliminate Type I errors and prevent false positives, one approach involves increasing the threshold for rejection to an extent that all medications are declined. Although this would effectively eradicate Type I errors, it would create an undesirable situation for patients.

On the other hand, to eliminate Type II errors and avoid overlooking beneficial drugs, a strategy of approving all medications could be considered. However, this approach is not widely endorsed due to its potential implications. Consequently, a trade-off between Type I and Type II errors is inevitable, prompting the need for a balanced approach.

The selection of 1.96 as a critical value is not arbitrary; it is based on extensive research across various fields, including biomedicine. The relative importance of Type I and Type II errors, as well as the weighting assigned to them, holds significant consequences.

This weighting is inherently linked to the alternative hypothesis, which posits the presence of an effect. In the context of this hypothesis, the probability of incorrectly rejecting a drug is influenced by the critical threshold and the distribution associated with the alternative hypothesis.

- Under alternative hypothesis H1: $\delta \equiv \mu_t - \mu_x = \kappa\sigma > 0$

$$n = 2\left(\frac{z_{\alpha/2} - z_{\beta}}{\kappa}\right)^2 \tag{9.4}$$

- Sample size n decreases with κ, increases with power $(1 - \beta)$

The trade-off between Type I and Type II errors becomes apparent through the consideration of the areas under the respective curves. This trade-off is quantified by the ratio of the areas, reflecting the relative significance of the two types of errors. In a simplified scenario, a trade-off with a ratio of approximately 12 suggests a preference for allowing multiple effective drugs to be approved rather than erroneously approving a single ineffective drug.

In addition to the consideration of Type I and Type II errors, the concept of statistical power, representing the likelihood of detecting an effective drug, is crucial. The power of a test, which is the complement of the Type II error, is a key indicator of a test's ability to identify an effect when it truly exists.

This power is influenced by the size of the treatment effect and the sample size, as well as the associated cost. Increasing the sample size can enhance the power of a test, thereby enabling a more precise distinction between the null and alternative hypotheses. However, the feasibility of such an approach is constrained by resource limitations, prompting the need to balance size, power, and cost when designing clinical trials.

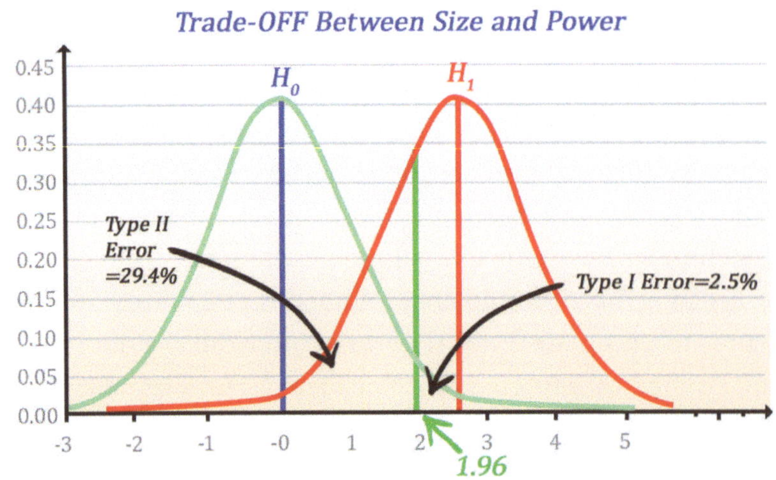

Figure 38: Trade-off between Size and Power

In the analysis of statistical data, a thorough examination beyond graphical representation is essential. The fundamental approach involves documenting all described parameters such as size, power, sample size, and others. Under the null hypothesis, it is possible to demonstrate that the probability of a Type I error is determined by , representing the probability of delta being greater than c. Subsequently, one can calculate the appropriate threshold to determine whether to reject the null hypothesis.

It is evident that this threshold decreases with the sample size, indicating that a larger patient population in a clinical trial allows for a more flexible approach in accepting or rejecting the null hypothesis due to the increased data accuracy. Furthermore, the threshold increases with the volatility of the underlying patient response variable and the specific statistical threshold.

Under the alternative hypothesis, it is feasible to compute the probability of a Type II error, which signifies the likelihood of not rejecting the null hypothesis under the alternative hypothesis. The alternative hypothesis assumes the existence of an effect, represented as a multiple of the standard deviation of the underlying patient response, where is a number greater than 0.

- Under alternative hypothesis H1: $\delta \equiv \mu_t - \mu_x = \kappa\sigma > 0$

$$\beta = \text{Prob}\ (\hat{\delta} \leq c)\ \text{(Type II error)}$$

$$= \text{Prob}\left(\frac{\hat{\delta} - \kappa\sigma}{\sqrt{\text{Var}[\hat{\delta}]}} \leq \frac{c - \kappa\sigma}{\sqrt{\text{Var}[\hat{\delta}]}}\right) = \text{Prob}\left(Z \leq \frac{c - \kappa\sigma}{\sigma\sqrt{2/n}}\right)$$

$$\beta = \Phi\left(\frac{c - \kappa\sigma}{\sigma\sqrt{2/n}}\right) = \Phi\left(\frac{\sigma z_{\alpha/2}\sqrt{2/n} - \kappa\sigma}{\sigma\sqrt{2/n}}\right)$$

$$= \Phi\left(z_{\alpha/2} - \frac{\kappa}{\sqrt{2/n}}\right)$$

$$\Phi^{-1}(\beta) = z_{\alpha/2} - \frac{\kappa}{\sqrt{2/n}} \Rightarrow z_\beta = z_{\alpha/2} - \frac{\kappa}{\sqrt{2/n}} \qquad (9.5)$$

The magnitude of κ determines the distance between the alternative hypothesis and the null, influencing the ease of identifying an effective drug during testing. Ultimately, this analysis determines the necessary sample size to achieve a specific power level. Typically, the industry seeks a power level ranging from 80% to 90%, corresponding to a Type II error rate of 10% to 20%, necessitating a sample size of approximately 1,000 to 5,000 patients. A reference table illustrates the implications for various power levels based on standard figures.

When discussing the effect of clinical trials on business deals, it is crucial to consider the standard randomized clinical trial, often characterized as a traditional two-armed balanced study with n patients in both the control and treatment groups. The primary objective is to compare the two groups to determine the treatment's impact on patients.

The focus of hypothesis testing revolves around the difference in treatment outcomes, with the null hypothesis assuming no difference and the alternative hypothesis suggesting a positive effect. Consequently, it is essential to navigate between these two hypotheses throughout the course of clinical trials, aiming to minimize errors while recognizing the inherent trade-offs.

In clinical trial design, the statistical framework of null and alternative distributions plays a pivotal role in decision-making. The null hypothesis assumes no effect of the drug, while the alternative hypothesis necessitates an assumption regarding the strength of the treatment effect. Balancing the probabilities of false positives and false negatives forms the crux of clinical trial design, influencing the crucial decision of patient enrollment.

Enrolling more patients enhances the likelihood of detecting deviations from the null hypothesis, albeit at a higher cost. The process involves calculating various statistical measures, including hypothesis tests and confidence intervals under the null hypothesis, to determine the trial's outcome. The 1.96 threshold serves as a critical criterion for approving or rejecting the null hypothesis in clinical trial analysis.

In the context of pharmaceutical and biotechnological enterprises, the focus lies on the alternative hypothesis, positing the existence of an effect. Consequently, the pivotal inquiry pertains to the magnitude of a clinical trial necessary to ensure the detection of the aforementioned effect. Previously, emphasis was placed on the probability of a Type II error, denoting the likelihood of a false negative. However, for pharmaceutical and biotechnological entities, it is imperative to minimize this probability.

Upon conducting an analysis and computing the Type II error, an expression emerges that delineates the inversion for the requisite number of patients to achieve a specific level of Type II error. Here, represents the alternative hypothesis, signifying the number of standard deviations by which the treatment effect is hypothesized under the alternative. Essentially, embodies the potency of the given treatment.

A higher value signifies a greater deviation from zero, indicating a more robust treatment effect. Consequently, a larger necessitates a smaller sample size to guarantee a certain Type I or Type II error, or a specific power.

In this context, denotes the probability of rejecting the null hypothesis, assuming it to be false. It can be likened to the likelihood of obtaining drug approval under the assumption of its efficacy.

Typically, clinical trials often utilize an 80% power threshold, although variations exist, with some trials employing 90% or slightly lower values based on circumstances. However, as a general rule, an 80% threshold is commonly employed.

The relationship between the number of patients required to achieve an 80% power and the specific alternative hypothesis is graphically represented. The graph illustrates the number of standard deviations by which the treatment deviates from a mean of 0, effectively indicating the efficacy of the therapy.

A more potent therapy, characterized by a substantial treatment effect such as life extension, blood pressure reduction, or tumor shrinkage, corresponds to a higher numerical value. Consequently, a larger treatment effect necessitates a smaller patient population to attain an 80% power or a 20% Type II error.

These numerical considerations have tangible financial implications. By multiplying the average cost of $36,000 per patient for a clinical trial by the required number of observations, the financial impact of the treatment effect strength becomes evident. A stronger treatment effect translates to a reduced patient requirement and subsequently lower expenditure.

For instance, a treatment effect of two standard deviations necessitates merely eight patients (four in each group), amounting to approximately $143,000 per arm or around $300,000 in total. Conversely, a smaller treatment effect, such as a tenth of a standard deviation, requires a substantial 1,500 patient trial, costing approximately $114 million for both arms. This illustrates how clinical trials can escalate in scale and cost based on the magnitude of the treatment effect pursued.

Ultimately, these considerations underscore the critical interplay between the strength of the treatment effect and the financial resources available. The strength of the treatment effect directly impacts the scale of risk, which in turn is contingent on available financing.

A weak treatment effect coupled with a small sample size entails substantial risk, whereas a strong treatment effect with a larger sample size signifies a more certain outcome. Consequently, the assessment and management of risk are intrinsically linked to financial considerations and the allocation of resources.

10. THE ETHICS OF BIOPHARMACEUTICAL PRICING AND VALUE

This chapter focuses on the interplay of pricing, value, and ethics within Biopharma and underscores the significance of ethics in influencing business decisions. Various influential factors drive decisions beyond mere financial considerations. Unlike the direct decision-making criteria of the NPV calculation, this chapter delves into intricate ethical dilemmas. The discourse may at times seem challenging as definitive conclusions may remain elusive. For instance, the ethical implications of pricing a drug in a manner that renders it unaffordable for some patients in dire need will be explored.

The chapter aims to spotlight these critical junctures, empowering readers to navigate business decisions impacting public health. The discussion commences with an exploration of the origins of ethics and their proper contemplation. Subsequently, the chapter addresses specific concerns, such as pricing considerations for cancer drugs. Finally, it delves into the distinction between price and value, and examines the application of cost-effectiveness analysis in shaping healthcare policies.

Ethically Determining Value in Healthcare

The biopharma industry is involved in the development and production of pharmaceutical drugs, biologics, and medical devices. It is characterized by high risk and the potential for substantial returns. The risks stem from the lengthy and costly process of drug development, stringent regulatory requirements, and the uncertainty of success in clinical trials. However, although successful drugs can yield significant returns through sales and licensing agreements, investors in this industry should carefully consider the ethical implications of their investments, weighing the potential for returns against the impact on public health and access to essential medications.

Economists often assess value in relation to market prices, which are determined by market dynamics. Products in high demand and abundant supply are typically priced lower than those in high demand but with limited supply. The existing pricing structures do not always align with logic or reason, but rather reflect the level of pricing that the market is willing to accept.

Price is therefore determined by what the market is willing to pay, and for economists, price and value are considered interchangeable. However, it is important to consider the ethical implications of pricing, particularly in the healthcare industry, where access to essential treatments and medications can have life-altering consequences for individuals and communities. Ethical considerations should be at the forefront of pricing decisions, ensuring that fair and equitable access to healthcare is maintained.

Ethics can be introduced to this context of human behavior by looking at an ethical dilemma involving a trolley car. In this scenario, a train is heading towards a platform with 10 people, and the brakes have failed, leaving the train unable to stop. The observer, a junior train assistant, can switch the train to a track where only one person is present. This situation presents a moral dilemma, as the observer must decide whether to allow the train to continue towards the 10 people or switch it to the track with one person.

The second moral dilemma involves the observer being on an overpass next to a stranger while the train is heading towards the 10 people. The question is posed whether the observer would be willing to push the stranger onto the tracks to slow down the train and save the 10 people.

This ethical analysis delves into the neurological and psychological aspects of decision-making in moral dilemmas. The first scenario, involving the train switch, triggers the prefrontal cortex responsible for logical deliberation, while the second scenario, involving pushing the stranger, triggers the mid-brain responsible for emotions and fear. This suggests that different parts of the brain are engaged when making decisions that involve actively harming or helping others.

The ethical implications of these moral dilemmas are complex and extend beyond simple utilitarian principles. The involvement of different parts of the brain in decision-making highlights the distinction between abstract calculation and emotional responses when considering actions that harm others. This discussion emphasizes the need to consider the ethical and moral implications of decision-making in various scenarios.

Carrying over to the context of Medicare and price controls, is the specific aspect of pricing within the domain of cancer treatment. For instance, in 2012, Memorial Sloan Kettering, a renowned cancer treatment center in New York City, made a contentious decision to decline the use of a particular cancer drug, Zaltrap, due to its exorbitant cost in comparison to existing drugs and its marginal benefits. This decision, considered unconventional, highlighted the ethical stance taken by the hospital against paying excessively for a drug that offered minimal advantages over existing options.

Additionally, is emphasis on the regulatory framework within which Medicare operates in relation to cancer drugs. There are constraints imposed on Medicare, particularly in the context of drugs covered under part B, which are administered at a physician's office. The US law mandates Medicare to cover any drug used in anti-cancer chemotherapeutics regimen, as long as it is for a medically accepted indication, regardless of the cost. This stands in contrast to the flexibility observed in the prescription and reimbursement of other treatments, where cost-effectiveness and generic options are typically considered.

Furthermore, there are issues of cancer drug shortages, particularly concerning sterile injectable drugs used in cancer treatment. The challenges faced by oncologists in accessing these drugs, leading to significant implications for patient care, including the need to switch chemotherapy regimens, delay treatments, or exclude patients from receiving necessary medications. The underlying reasons for these shortages, as documented in a study by the Department of Health and Human Services, point to capacity constraints and market dynamics, wherein manufacturers respond to incentives, leading to fluctuations in availability and pricing.

There is therefore a complex interplay between regulatory frameworks, market dynamics, and ethical considerations within the domain of cancer drug pricing and availability, emphasizing the multifaceted nature of the challenges faced in this critical area of healthcare.

The Distinct Pricing Nature of Medicine

The differentiation of medicine from other commodities lies in the disparity between value and pricing. In the realm of medicine, consumers lack the authority to make decisions. In contrast, when purchasing a car, a consumer has the autonomy to decide whether or not to acquire a specific model based on personal preference.

However, this direct consumer decision-making is absent in the context of acquiring a cancer drug. In such critical medical scenarios, medical professionals are the decision-makers and are not focused on cost control or prescribing based on a patient's financial capacity. The expectation is for medical professionals to administer the best possible treatment, irrespective of cost.

Furthermore, the intricate nature of the healthcare system complicates matters, as the actual cost of a particular treatment remains opaque due to multiple layers of payments and cost sharing. Consequently, the necessity for data to make informed economic decisions becomes apparent. However, the absence of measurable data poses a challenge. These factors collectively contribute to the unique nature of medicine.

The idea of establishing a "Consumer Reports" equivalent for medical treatments prompts the question of its viability. Such a system would enable individuals to access and evaluate various treatment records and their value, akin to assessing the performance and value of different car models. However, the mere existence of such a system does not universally influence decision-making, as evidenced by personal experiences.

Yet, if a method to measure both cost and benefit, such as quality-adjusted life years, were available, individuals could make informed decisions about the cost-effectiveness of treatments. One such measure, known as the incremental cost-effectiveness ratio (ICER), calculates the change in cost divided by the change in health, determining the additional health benefits obtained from a new therapy that costs more than a previous one.

This approach, along with various methodologies in value-based pricing, health technology assessment, and Pharmacoeconomics, aims to assess the value of treatments across different therapeutic areas objectively. The purpose of these methods is to discern when to decline a particular treatment. If the decision to decline is never necessary, then such evaluations become redundant.

Therefore, the willingness to refuse certain treatments becomes a critical aspect of this process. Assessing the value-addition of using different endpoints for clinical trials, as well as the intricate analysis of cost-benefit, forms a complex but formal conclusion, reliant on a series of assumptions. This exercise in Pharmacoeconomics is considered controversial and subject to criticism, as it is challenging to achieve perfection in cost-benefit analysis due to the complexity of the task.

The United Kingdom houses an institution known as NICE, the National Institute for Health and Care Excellence, which offers assessments regarding the appropriateness of specific therapies based on their cost and indications. NICE functions as an advisory body to the National Health Service (NHS), which holds the ultimate responsibility for decision-making. The organization's website contains comprehensive guidelines, reflecting their conscientious approach to evaluating the cost-effectiveness and benefits of various treatments.

NICE employs social value judgment principles and bioethical considerations, encompassing utilitarian and egalitarian arguments, to arrive at decisions that uphold procedural justice and transparency. Their evaluation of Sovaldi, for instance, delineates specific indications and combinations for which the drug is recommended or not, along with detailed justifications for their determinations.

This thorough process is encapsulated in a 150-page report, openly presenting the analysis and rationale behind their recommendations. With a commitment to transparency, NICE provides a flowchart on their website, allowing stakeholders to understand the permitted treatments for specific conditions and the underlying reasoning.

However, NICE's decision-making has sparked controversy, particularly when evaluating Kadcyla, a breast cancer drug. Despite its ability to extend life by approximately nine months, NICE declined to recommend it due to its high annual cost of 90,000 pounds. This decision elicited public outcry, with some questioning the ethical implications of denying access to potentially life-prolonging treatments.

The debate surrounding such decisions underscores the complex interplay between healthcare costs, individual well-being, and societal considerations. The discourse also raises questions about the ethical and practical implications of prioritizing healthcare resources.

The challenges of such decisions are further highlighted when considering the broader impact on healthcare costs and access. While these deliberations may appear straightforward when framed in personal terms, they have far-reaching implications for healthcare systems, affecting the affordability and accessibility of treatments for all individuals.

This underscores the need for a comprehensive and nuanced approach to decision-making in healthcare policy, one that considers the broader implications for society as a whole. This also necessitates a careful balance between considerations of the greater good and the equitable distribution of healthcare resources, a task that demands thoughtful and responsible leadership.

The Bottom Line

In the realm of healthcare economics, better strategies are essential to usher in novel approaches, given the trajectory of current healthcare trends. Data indicates that the nation allocated 17% of its Gross Domestic Product (GDP) to healthcare. By contrast, the red line illustrates the United Kingdom's healthcare spending. Despite the differing levels and associated complexities due to the UK's smaller size, the crucial point remains: the UK has successfully altered its trajectory to a downward slope. This achievement may be linked to the actions of the NICE.

Renowned investor Warren Buffett recently emphasized that the looming challenge does not stem from taxes, but rather from escalating healthcare costs. While acknowledging that expending 17% of the US GDP on healthcare seems exorbitant, affluence and the tangible returns are derived from such investments. Contrary to popular belief, the US could feasibly allocate 35% of its GDP to healthcare, not as an endorsement, but as a reflection of the nation's financial capacity.

For instance, an introspective look at the financial sacrifices one is willing to make if faced with a loved one's life-threatening illness underscores the dilemma posed by the burgeoning opportunities in medical science, which inevitably lead to increased healthcare expenditure. The concern therefore lies not in the current expenditure levels, but in the realization that economies can afford to allocate even more resources to healthcare.

Efficiency in healthcare spending across nations, particularly in comparison to outcomes, is another crucial consideration. Transitioning to a single-payer system could potentially improve economic outcomes and lower costs. However, the pharmaceutical industry argues that the US, as the largest market for new drugs, gains access to a wider array of medications, making it a desirable healthcare destination for many.

The development of groundbreaking therapies, such as gene therapy for blindness by Spark Therapeutics, presents ethical and financial dilemmas. The ability to afford such treatments raises concerns about the rational allocation of resources and the ethical implications of choosing who to help and who to not.

Proposed business models involve a last-resort clinic catering to high-net-worth individuals and employing experimental therapies. The unique fee structure, driven by hedge fund principles, aims to incentivize the development of groundbreaking treatments, which would ultimately be offered to the broader population at no cost. While legal and ethical hurdles currently impede the realization of this proposal, it presents a potential approach to addressing the challenges of healthcare economics in a new and innovative manner. This necessitates a reevaluation of ethical considerations and decision-making processes.

BIBLIOGRAPHY

Adleman, L. M. (1994). Molecular computation of solutions to combinatorial problems. Science.

Aggarwal, S. (2014). What's fueling the biotech engine-2012 to 2013. Nature Biotechnology, 32(1), 32-39.

Anatol, R., et al. (2014). Chapter 86 - The Regulatory Process from Concept to Market. In R. Lanza, R. Langer, & J. Vacanti (Eds.), Principles of Tissue Engineering (Fourth Edition) (pp. 1781-1797). Academic Press. ISBN 978-0-12-398358-9. doi: 10.1016/B978-0-12-398358-9.00086-0.

Anthony, J. M., Haworth, P., & Mills, G. M. (2020). Biotechnology business development: The art of the deal. In C. Shimasaki (Ed.), Biotechnology Entrepreneurship (2nd ed., pp. 541-551). Academic Press. https://doi.org/10.1016/B978-0-12-815585-1.00036-X.

Arrow, K. J. (1962). Economic welfare and the allocation of resources for invention. In R. R. Nelson (Ed.), The rate and direction of inventive activity: Economic and social factors (pp. 609-626). Princeton University Press.

Arthritis Foundation. (n.d.). Biologics. Arthritis Foundation. Retrieved from https://www.arthritis.org/drug-guide/biologics/biologics

Bansal, S., Mahendiratta, S., Kumar, S., Sarma, P., Prakash, A., & Medhi, B. (2019). Collaborative research in modern era: Need and challenges. Indian Journal of Pharmacology, 51(3), 137-137. https://doi.org/10.4103/ijp.ijp_394_19.

BioCentury. (n.d.). Financial dashboard [Webpage]. BioCentury. Retrieved from https://www.biocentury.com/bciq-data/financial-dashboard?id=cku8n0acwk1ub0b17yla1ki9p&item=cku8n3ivcks2g0c28ji87m0c9

Biotechnology Innovation Organization. (n.d.). How do drugs and biologics differ? Biotechnology Innovation Organization. Retrieved from https://archive.bio.org/articles/how-do-drugs-and-biologics-differ

Brealey, R. A., Myers, S. C., & Allen, F. (2006). Principles of Corporate Finance. McGraw-Hill/Irwin.

Callaway, E. (2019). 'It will change everything': DeepMind's AI makes gigantic leap in solving protein structures. Nature. https://doi.org/10.1038/d41586-019-01400-6

Campen, L. E. V., Poplazarova, T., Therasse, D. G., & Turik, M. (2021). Considerations for applying bioethics norms to a biopharmaceutical industry setting. BMC Medical Ethics, 22(1). https://doi.org/10.1186/s12910-021-00600-y.

Carden, C., Chamberlain, T., & Hill, J. W. (2010). The brave new world of valuing life sciences and healthcare enterprises. Business Horizons, 53(2), 183-197. https://doi.org/10.1016/j.bushor.2009.11.004.

Caring Info. (n.d.). Curative care.

CaringInfo. Retrieved from https://www.caringinfo.org/types-of-care/curative-care/#:~:text=Curative%20or%20therapeutic%20care%20refers,a%20cure%20is%20not%20possible

Carter, P. H., Berndt, E. R., DiMasi, J. A., & Trusheim, M. R. (2016). Investigating investment in biopharmaceutical R&D. Nature Reviews Drug Discovery, 15(10), 673-674. https://doi.org/10.1038/nrd.2016.104.

Church, G. M., Gao, Y., & Kosuri, S. (2012). Next-generation digital information storage in DNA. Science.

CNN. (2022, February 2). Cancer patients in remission years after T-cell therapy, study says. CNN. Retrieved from https://edition.cnn.com/2022/02/02/health/cancer-t-cell-therapy-remission-study/index.html

Coventry. (n.d.). Understanding the difference between specialty drugs, biologics, and biosimilars. Coventry. Retrieved from https://www.coventrywcs.com/insights/workers-comp/blog/difference-between-specialty-drugs-biologics-and-biosimilars

Cutler, D., Kirson, N., & Long, G. (2020). Financing Drug Innovation in the US: Current Framework and Emerging Challenges. PharmacoEconomics, 38(9), 905–911. https://doi.org/10.1007/s40273-020-00926-2.

Cutting Edge Information. (n.d.). FMV Connect Pro and FMV Data. Cutting Edge Information. Retrieved from https://cuttingedgeinfo.com/fmvconnect-pro-and-fmv-data/

Damodaran, A. (2016). Investment Valuation: Tools and Techniques for Determining the Value of Any Asset. Wiley.

Data, J. (2017). Chapter 42 - Pharmaceutical and Biotechnology Sector Support of Research. In D. Robertson & G. H. Williams (Eds.), Clinical and Translational Science (Second Edition) (pp. 745-752). Academic Press. ISBN 978-0-12-802101-9. doi: 10.1016/B978-0-12-802101-9.00042-9.

Davis, A. M., Engkvist, O., Fairclough, R. J., Feierberg, I., Freeman, A., & Iyer, P. (2021). Public-private partnerships: Compound and data sharing in drug discovery and development. SLAS Discovery, 26(5), 604-619. DOI: 10.1177/2472555220982268.

Deloitte. (2024, February 9). Life Sciences | DART – Deloitte Accounting Research Tool. https://dart.deloitte.com/USDART/home/publications/deloitte/industry/life-sciences

DiMasi, J. A., Feldman, L., Seckler, A., & Wilson, A. (2010). Trends in risks associated with new drug development: success rates for investigational drugs. Clinical Pharmacology & Therapeutics, 87(3), 272-277.

DiMasi, J. A., Grabowski, H. G., & Hansen, R. W. (2016). Innovation in the pharmaceutical industry: New estimates of R&D costs. Journal of Health Economics, 47, 20-33.

Doudna, J. A., & Charpentier, E. (2014). The new frontier of genome engineering with CRISPR-Cas9. Science, 346(6213).

Driscoll, D., Farnia, S., Kefalas, P., & Maziarz, R. T. (2017). Concise review: The high cost of high tech medicine: Planning ahead for market access. Stem Cells Translational Medicine, 6(8), 1723-1729. https://doi.org/10.1002/sctm.16-0487. Donoghue, J. P. (2002). Connecting cortex to machines: recent advances in brain interfaces. Nature Neuroscience.

European Medicines Agency. (n.d.). Orphan designation. European Medicines Agency. Retrieved from https://www.ema.europa.eu/en/glossary/orphan-designation#:~:text=A%20status%20assigned%20to%20a,competition%20once%20on%20the%20market

Flier, J. S. (2019). Academia and industry: Allocating credit for discovery and development of new therapies. Journal of Clinical Investigation, 129(6), 2172-2174. https://doi.org/10.1172/jci129122.

Ford, D., & Nelsen, B. (2014). The view beyond venture capital. Nature Biotechnology, 32(1), 15-23. https://doi.org/10.1038/nbt.2780.

Freedman, T. (2014). Chapter 31 - Career Opportunities in the Life Sciences Industry. In C. Shimasaki (Ed.), Biotechnology Entrepreneurship (pp. 435-444). Academic Press. ISBN 9780124047303. https://doi.org/10.1016/B978-0-12-404730-3.00031-2.

Germann, P. G., Schuhmacher, A., Harrison, J., Law, R. E., Haug, K., & Wong, G. (2013). How to create innovation by building the translation bridge from basic research into medicinal drugs: An industrial perspective. Human Genomics, 7(1). https://doi.org/10.1186/1479-7364-7-5.

Gersdorf, T., He, V. F., Schlesinger, A., Koch, G., Ehrismann, D., Widmer, H., & von Krogh, G. (2019). Demystifying industry-academia collaboration. Nature reviews. Drug discovery, 18(10), 743–744. https://doi.org/10.1038/d41573-019-00001-2.

Gleadle, P., Parris, S., Shipley, M., & Simonetti, R. (2014). Financing Pharmaceutical Innovation. Journal of Financial Transformation.

Hamburg, M. A., & Collins, F. S. (2010). The path to personalized medicine. New England Journal of Medicine, 363(4), 301-304.

Herper, M. (2013). FDA approves first 'bionic eye'. Forbes. https://www.forbes.com/sites/matthewherper/2013/02/14/fda-approves-first-bionic-eye/?sh=4dbcbcd32667

High, K. A., & Roncarolo, M. G. (2019). Gene Therapy. N Engl J Med.

Honig, P. K., & Lalonde, R. L. (2010). The economics of drug development: A grim reality and a role for clinical pharmacology. Clinical Pharmacology & Therapeutics, 87(3), 247-251. https://doi.org/10.1038/clpt.2009.298.

Hopkins, M. M., Martin, P. A., Nightingale, P., Kraft, A., & Mahdi, S. (2007). The myth of the biotech revolution: An assessment of technological, clinical and organisational change. Research Policy, 36(4), 566-589.

Hourd, P. C., & Williams, D. J. (2006). Success in healthcare technology businesses: Coordinating the value milestones of new product introduction, financial stakeholders and business growth. Innovation: Management, Policy and Practice, 8(3), 229-247. https://doi.org/10.5172/impp.2006.8.3.229.

Humphrey, A. E. (1996). Some issues in biotechnology commercialization. Technology in Society, 18(3), 321-332. DOI: 10.1016/0160-791x(96)00017-6.

In Vivo. (2024, February 9). New Payment and Financing Models For Curative Regenerative Medicines. https://invivo.citeline.com/IV005132/New-Payment-And-Financing-Models-For-Curative-Regenerative-Medicines

Informa Connect. (2024, February 9). Home | Finance & Accounting, Life Sciences. https://informaconnect.com/finance-accounting-life-sciences

Investopedia. (n.d.). Follow-on offering. Investopedia. Retrieved from https://www.investopedia.com/terms/f/followonoffering.asp#:~:text=What%20Is%20Follow%2Don%20Financing,start%20up%20has%20gone%20public

Investopedia. (n.d.). Private investment in public equity (PIPE). Investopedia. Retrieved from https://www.investopedia.com/terms/p/pipe.asp

Johnson, A. (2022). CHS freshman doesn't let near-blindness slow him down. Arkansas Catholic. Retrieved from https://www.arkansas-catholic.org/news/article/6960/CHS-freshman-doesnt-let-near-blindness-slow-him-down

Kaitin, K. I. (2010). Deconstructing the drug development process: the new face of innovation. Clinical Pharmacology & Therapeutics, 87(3), 356-361.

Kasatova, A. A., Vagizova, B., & Kokh, I. A. (2020). Financial potential and tools for commercialization of biotechnology projects in the sustainable development system. International Journal of Engineering Research and Technology, 13(11), 3659-3666. DOI: 10.37624/ijert/13.11.2020.3659-3666.

Kessel, M., & Frank, F. C. (2007). A better prescription for drug-development financing. Nature Biotechnology, 25(8), 859-866. https://doi.org/10.1038/nbt0807-859.

Kickbusch, I., Krech, R., Franz, C., & Wells, N. (2018). Banking for health: Opportunities in cooperation between banking and health applying innovation from other sectors. BMJ Global Health, 3(Suppl 1), e000598. https://doi.org/10.1136/bmjgh-2017-000598.

Kleiman, R. J., & Ehlers, M. (2019). How to develop therapeutic and translational research collaborations with industry. Molecular Biology of the Cell, 30(22), 2741-2743. https://doi.org/10.1091/mbc.e19-05-0261.

Kola, I., & Landis, J. (2004). Can the pharmaceutical industry reduce attrition rates? Nature Reviews Drug Discovery.

Krech, R., Kickbusch, I., Franz, C., & Wells, N. (2018). Banking for health: The role of financial sector actors in investing in global health. BMJ Global Health, 3(Suppl 1), e000597. https://doi.org/10.1136/bmjgh-2017-000597.

Lanza, R., & Atala, A. (2013). Handbook of Stem Cells. Elsevier.

Law Insider. (n.d.). Definition of preventive therapy. Law Insider. Retrieved from https://www.lawinsider.com/dictionary/preventive-therapy

Lederberg, J., & McCray, A. T. (2001). 'Ome Sweet 'Omics— A Genealogical Treasury of Words. The Scientist, 15(7).

Lo, A. W., & Chaudhuri, S. E. (2022). Healthcare Finance: Modern Financial Analysis for Accelerating Biomedical Innovation [Ebook]. Princeton University Press.

Mayo Clinic. (n.d.). Monoclonal antibody drugs for cancer: How they work. Mayo Clinic. Retrieved from https://www.mayoclinic.org/diseases-conditions/cancer/in-depth/monoclonal-antibody/art-20047808

McKinsey & Company. (n.d.). How the European biotech sector can navigate turbulent times. McKinsey & Company. Retrieved from https://www.mckinsey.com/industries/life-sciences/our-insights/how-the-european-biotech-sector-can-navigate-turbulent-times#/

McKinsey & Company. (n.d.). The bio-revolution: Innovations transforming economies, societies, and our lives. McKinsey & Company. Retrieved from https://www.mckinsey.com/industries/life-sciences/our-insights/the-bio-revolution-innovations-transforming-economies-societies-and-our-lives

Mohs, R. C., & Greig, N. H. (2017). Drug discovery and development: Role of basic biological research. Alzheimer's & Dementia: Translational Research & Clinical Interventions, 3(4), 651-657. https://doi.org/10.1016/j.trci.2017.10.005

Munos, B. (2009). Lessons from 60 years of pharmaceutical innovation. Nature Reviews Drug Discovery.

Myers, S. C., & Majluf, N. S. (1984). Corporate financing and investment decisions when firms have information that investors do not have. Journal of Financial Economics, 13(2), 187-221.

Naldini, L. (2015). Gene therapy returns to centre stage. Nature.

National Center for Biotechnology Information. (2015). Clinical trials: The evolution and future of Phase 1 trials. NCBI. Retrieved from https://www.ncbi.nlm.nih.gov/pmc/articles/PMC4467013/

National Institutes of Health (NIH). (2020). Funding for Biomedical Research.

National Institutes of Health (NIH). (2020). Stem Cell Information.

National Organization for Rare Disorders. (2018). Orphan diseases: Myths & facts [Infographic]. Retrieved from https://rarediseases.org/wp-content/uploads/2018/05/NRD-1123-5_OD_Myths_Infographic_FNL2.pdf

Neelapu, S. S., Locke, F. L., Bartlett, N. L., et al. (2017). Axicabtagene ciloleucel CAR T-cell therapy in refractory large B-cell lymphoma. New England Journal of Medicine, 377(26), 2531-2544.

News Medical. (n.d.). Drug patents and generics. News-Medical.net. Retrieved from https://www.news-medical.net/health/Drug-Patents-and-Generics.aspx

O'Brien, S. (2016, May 27). Spark Therapeutics' gene therapy offering fresh hope to blind patients. CNBC. Retrieved from https://www.cnbc.com/2016/05/27/spark-therapeutics-gene-therapy-offering-fresh-hope-to-blind-patients.html

Organisation for Economic Co-operation and Development (OECD). (n.d.). Pharmaceutical pricing policy. OECD. Retrieved from https://www.oecd.org/els/health-systems/pharmaceutical-pricing-policy.htm

Pammolli, F., Magazzini, L., & Riccaboni, M. (2011). The productivity crisis in pharmaceutical R&D. Nature Reviews Drug Discovery, 10(6), 428-438.

Paul, S. M., Mytelka, D. S., Dunwiddie, C. T., Persinger, C. C., Munos, B. H., Lindborg, S. R., & Schacht, A. L. (2010). How to improve R&D productivity: the pharmaceutical industry's grand challenge. Nature Reviews Drug Discovery, 9(3), 203-214.

Pharmaceutical Research and Manufacturers of America (PhRMA). (2020). Biopharmaceutical R&D: The Process Behind New Medicines.

Pisano, G. P. (2006). Can science be a business? Lessons from biotech. Harvard Business Review, 84(10), 114-125.

Pisano, G. P. (2006). Science Business: The Promise, the Reality, and the Future of Biotech. Harvard Business Press.

Popa, C., Holvoet, K., Montfort, T. V., Groeneveld, F., & Simoens, S. (2018). Risk-return analysis of the biopharmaceutical industry as compared to other industries. Frontiers in Pharmacology, 9. https://doi.org/10.3389/fphar.2018.01108.

PricewaterhouseCoopers LLP (PwC). (n.d.). Six drug pricing models: Which ones are right for your products? PricewaterhouseCoopers LLP (PwC). Retrieved from https://wwwStatista. (n.d.). Global R&D expenditure for pharmaceuticals. Statista. Retrieved from https://www.statista.com/statistics/309466/global-r-and-d-expenditure-for-pharmaceuticals/.pwc.com/us/en/industries/health-industries/library/6-drug-pricing-models.html

Quinn, C., Young, C. M., Thomas, J. C., & Trusheim, M. R. (2019). Estimating the clinical pipeline of cell and gene therapies and their potential economic impact on the US healthcare system. Value in Health, 22(6), 621-626. https://doi.org/10.1016/j.jval.2019.03.014.

Rai, M., & Goyal, R. (2018). Chapter 33 - Pharmacoeconomics in Healthcare. In D. Vohora & G. Singh (Eds.), Pharmaceutical Medicine and Translational Clinical Research (pp. 465-472). Academic Press. ISBN 978-0-12-802103-3. doi: 10.1016/B978-0-12-802103-3.00034-1.

Ritter, J. R. (2003). Investment banking and securities issuance. In R. A. Jarrow, V. Maksimovic, & W.

T. Ziemba (Eds.), Handbook of the Economics of Finance (pp. 255-306). Elsevier.

Rosen, J., Hannaford, B., & Satava, R. M. (2014). Surgical Robotics: Systems Applications and Visions. Springer.

Russell, S., Bennett, J., Wellman, J. A., et al. (2017). Efficacy and safety of voretigene neparvovec (AAV2-hRPE65v2) in patients with RPE65-mediated inherited retinal dystrophy: a randomised, controlled, open-label, phase 3 trial. Lancet, 390(10097), 849-860.

Rx Benefits. (n.d.). What are orphan drugs, biologics, and biosimilars? *Rx Benefits*. Retrieved from https://www.rxbenefits.com/blogs/what-are-orphan-drugs-biologics-and-biosimilars/#:~:text=An%20emerging%20sub%2Dclass%20of,of%20200%2C000%20or%20fewer%20people

Sammut, S. M. (2020). Chapter 21 - Financial Ramifications of Funding a Biotechnology Venture: What You Need to Know About Valuation and Term Sheets. In C. Shimasaki (Ed.), Biotechnology Entrepreneurship (Second Edition) (pp. 299-312). Academic Press. ISBN 978-0-12-815585-1. doi: 10.1016/B978-0-12-815585-1.00021-8.

Saur-Amaral, I., & Gouveia, J. B. (2007). Uncertainty in drug development: Insights from a Portuguese firm. International Journal of Technology Intelligence and Planning, 3(4), 355-355. https://doi.org/10.1504/ijtip.2007.016306.

Second Sight Medical Products. (2013). Argus II Retinal Prosthesis System.

Senior, M. (2023). Precision financing. Nature Biotechnology, 41(5), 597-603. https://doi.org/10.1038/d41587-023-00001-z. Markowitz, H. (1952). Portfolio selection. The Journal of Finance, 7(1), 77-91.

Sharma, A., Amarnath, S., & Rath, G. (2019). Gene Therapy and its Application in Musculoskeletal Disorders. Journal of Clinical Orthopaedics and Trauma, 10(1), 9–15. https://doi.org/10.1016/j.jcot.2018.12.012

Siew, M. S. (n.d.). Biotechnology vs Biomedical Science: What's the Difference? EduAdvisor. Retrieved from https://eduadvisor.my/articles/biotechnology-vs-biomedical-science-whats-the-difference

Silverman, E. (2004). Follow the money: How biotech's funding shapes its direction. PubMed.Gassmann, O., Reepmeyer, G., & Zedtwitz, M. (2008). Leading pharmaceutical innovation: trends and drivers for growth in the pharmaceutical industry. Springer Science & Business Media.

Smith, D. G., Culme-Seymour, E., & Mason, C. (2018). Evolving industry partnerships and investments in cell and gene therapies. Cell Stem Cell, 22(5), 623-626. https://doi.org/10.1016/j.stem.2018.03.004.

Tageja, N. (2010). Bridging the translation gap – new hopes, new challenges. Journal of Clinical Pharmacy and Therapeutics, 35(2), 163-171. https://doi.org/10.1111/j.1472-8206.2010.00903.x

Truveris. (n.d.). Understanding the difference between specialty drugs and traditional medications. Truveris. Retrieved from https://truveris.com/difference-between-specialty-drug/

U.S. Center for Medicare & Medicaid Services (n.d.). Brand-name drugs. HealthCare.gov. Retrieved from https://www.healthcare.gov/glossary/brand-name-drugs/

U.S. Department of Health & Human Services (HHS). (n.d.). Medicaid drug pricing summary [PDF file]. Retrieved from https://www.hhs.gov/sites/default/files/HCVMedicaidDrugPricingSummary.pdf

U.S. Food and Drug Administration. (2018). Biologics Research and Development.

U.S. Food and Drug Administration. (n.d.). Exclusivity and Generic Drugs: What Does It Mean? [PDF file]. Retrieved from https://www.fda.gov/files/drugs/published/Exclusivity-and-Generic-Drugs--What-Does-It-Mean-.pdf

Wanerman, R. E., Javitt, G. H., & Shah, A. B. (2020). Chapter 29 - Artificial Intelligence in Biotechnology: A Framework for Commercialization. In C. Shimasaki (Ed.), Biotechnology Entrepreneurship (Second Edition) (pp. 419-427). Academic Press. ISBN 978-0-12-815585-1. doi: 10.1016/B978-0-12-815585-1.00029-2.

Xu, B., Magnan, M.L., & André, P.E. (2007). The Stock Market Valuation of R&D Information in Biotech Firms. Contemporary Accounting Research, 24, 1291-1318. https://doi.org/10.1506/car.24.4.10

Yan, S. (2017). Selected Concepts on Ethics in Biopharmaceutical Innovation. MOJ Proteomics & Bioinformatics, 6(4), 10.15406/mojpb.2017.06.00204.

Bei Fragen zur Produktsicherheit wenden Sie sich bitte an:
If you have any questions regarding product safety,
please contact:

Walter de Gruyter GmbH
Genthiner Straße 13
10785 Berlin
productsafety@degruyterbrill.com